Liaison Functions 2020

Held at the 2020 AIChE Spring Meeting and 16th Global Congress on Process Safety

Online
17-21 August 2020

ISBN: 978-1-7138-2347-6

Printed from e-media with permission by:

Curran Associates, Inc.
57 Morehouse Lane
Red Hook, NY 12571

AIChE
120 Wall Street, FL 23
New York, NY 10005-4020

Phone: (800) 242-4363
Fax: (203) 775-5177

www.aiche.org

Additional copies of this publication are available from:

Curran Associates, Inc.
57 Morehouse Lane
Red Hook, NY 12571 USA
Phone: 845-758-0400
Fax: 845-758-2633
Email: curran@proceedings.com
Web: www.proceedings.com

Liaison Functions 2020

Held at the 2020 AIChE Spring Meeting and 16th Global Congress on Process Safety

Online
17-21 August 2020

AIChE
The Global Home of Chemical Engineers

TABLE OF CONTENTS

591161 2020 AIChE Spring Meeting & 16th GCPS AGILE Award Keynote Address - Beyond the Horizon: The Factors and Forces Affecting the Global Petrochemical Industry

Abstract:

Over the past 10 years, LyondellBasell and the global petrochemical industry has undergone a dramatic transformation. In this era of unprecedented disruption, it is likely that the next the decade will present both challenges and opportunities for the petrochemical sector. LyondellBasell CEO Bob Patel will summarize the company's progress to date and provide a forward-looking perspective on key opportunities and risks faced by the industry in 2020 and beyond.

614720 Does Age Really Matter?

Abstract:

The keynote, "Does Age Really Matter" – will highlight thoughts on two very critical assets in the industry. First, Manufacturing Facilities that are aging and experiencing significant incidents, and second, aging of our human resource and significant loss of critical process safety management knowledge, and their ultimate impact on the process safety management performance. Thoughts on what industry and other organizations are doing to address these gaps would be reviewed.

592550 Digital Transformation in Chemical Industry

Abstract:

Manufacturing is the original denizen of one of the cornerstones of Industry 4.0 - a digital transformation journey to unlock the full potential of manufacturing data. This talk will highlight the critical human element to deliver sustainable and scalable digital transformation, discuss common challenges using Artificial Intelligence and Advanced Analytics as an example of digital transformation, focusing on corporate culture change. A proven best practice will be shared to illustrate real-world outcome in digital transformation.

615001 Frontlines of COVID-19: Lead, Learn and Cure

Abstract:

A diversified expert panel will share their experience in managing COVID-19 risks through applying robust corporate leadership and culture, transforming incidents investigation tech.niques and learning from recent medicinal development. The panel will highlight the role of the scientific community to manage the COVID-19 Risk

591157 Securing Enterprises via Cybersecurity Understanding, Resiliency, and Education

Abstract:

Securing Enterprises via Cybersecurity Understanding, Resiliency, and Education

Industry 4.0 initiatives are enabled by leveraging the deployment of information technologies into the operational technology environment. While a logical approach in defining the future of smart manufacturing, it remains fret with challenges and risks that must be mitigated assure enterprise resilience to cyber and cyber-physical attacks. As economies recalibrate in the aftermath of the current public health emergency, definitions of what constitutes "critical infrastructure" are broadening while, at the same time, attack surfaces and the threat landscape are expanding. This means companies must place a greater emphasis on taking the steps and making the investments necessary to overcome the challenges and navigate these challenges and risks to maximize their resilience and secure and protect their enterprises. This presentation will explore the link between cybersecurity and process safety in the context of a safety culture. It will also identify cybersecurity challenges in industry 4.0 initiatives, the integration of information technologies (IT) and operational technologies (IT), the delivery of training, and the need for credentialing and ongoing professional development.

612627 Management and Allyship of Underrepresented Groups

Abstract:

Managing an increasingly diverse and global workforce is itself a challenging skill to improve upon, but what best practices can be used when specifically leading, cultivating, and growing teams that represent underrepresented groups in gender, minorities, and LGBTQIA+? Not only is a diverse workforce the new global norm in modern business, but having diverse teams has also been linked to higher innovation revenues and market growth in studies by McKinsey and Boston Consulting Group. This interactive hands-on workshop will involve discussions of real world experiences of managers and talent development leads on what it means to be an ally, how to identify inclusion gaps in your organization, and how to demonstrate best practices in world class leadership.

590220 Global Ethylene/Propylene: Refinery Scale Petrochemical Investments and the Impact on Global Markets

Abstract:

Previous announcements have become a reality concerning new refineries built on a chemicals platform with fuel as the byproduct. Long term these type of investments represent a trend that could truly change the market balance. In this presentation we will examine the differing motivations and technologies behind refinery scale olefin investments, how does North America fit into the supply/demand picture, and other long term factors that will play into petrochemical investment decisions.

590229 Disruptive Trends and their Implications

Abstract:

Braskem, known as a leading innovator in petrochemical sustainability, also has a robust transformation agenda preparing for megatrends that could reshape the industry as we know it. In his talk, Kevin will share some real-world examples of exponential change and how disruptive it can be to the most well-established of industries. Exponential change is all around us, but we as humans are often terrible at perceiving and forecasting that change until large disruptions occur. Kevin will talk through four key megatrends impacting the petrochemical industry – Mobility, Energy, Digitization, and Sustainability – and give some examples of how these areas could be disrupted and how Braskem is preparing and embracing these changes.

591208 How to Pivot In Your Career

Abstract:

Pivoting your career and changing course in the middle of your career may be a bit scary, but it can also be strategic. Whether you are shifting employers, taking on new tasks, moving to a new city, or changing careers or industries altogether, there are a number of steps you can take to ensure that you achieve your goals in an efficient and satisfactory manner. We will focus on developing the strategy and employing the tactics to find an opportunity that is right for you, go after it, complete a smooth transition, and maintain a stellar reputation in and after the process. Let's prepare you for your new adventure!

591534 Engineers at the Wheel

Abstract:

The multiple challenges faced in our industry require a combination of solid analytical and technical skills with a versatile leadership style and passion for helping people. This presentation will cover learnings, successes and challenges from engineers that shaped the chemical industry in the last century and will discuss how their lessons can apply to challenges faced by Chemical Engineers today.

592220 Leadership in Crisis

Abstract:

Leadership in time of crisis. How to lead an organization to plan for worst case scenarios (preoccupation with failure) and then how to lead an organization when the crisis event occurs. This talk will bring in relevant scenarios from major events like the response to Harvey and other unplanned events impacting the industry. It will also share some thoughts on the need to consider worst case scenarios, emergency preparedness, expecting the unexpected, during our technical work as engineers.

592593 Trailblazing in the Energy Industry

Abstract:

In this talk, Fatimah Bello, VP, Business Development & Product at Heliogen will share important lessons from leadership from her transition from traditional chemical engineering roles to leading disruption in the energy field through the implementation of new solar based technologies.

589775 Data Infrastructure As Foundation to Operational Excellence and Digital Transformation

Abstract:

Is technology a solution looking for a problem? A successful digital transformation builds on the foundations of Operational Excellence. Innovations in the technology we use further these initiatives. The building blocks of these improvements are rooted in the data we collect, compile, and contextualize into operational intelligence. Successful implementations identify desired outcomes and build the data and technical infrastructures to achieve them.

A structured Operational Excellence program begins with clear definitions for:

- What's important?
- Who's responsible for action, who's accountable for success?
- How is success measured?
- How are excursions identified, monitored, and corrected?
- What communications are necessary to ensure improvement?
- What skills and tools are necessary to continuously improve?

Compelling events for transformation may include regulatory compliance, hazard recognition, sustainability, cost controls, quality improvements, and asset reliability as examples. The data required covers the spectrum of real-time operational data, environmental monitoring, quality data, and asset information. A connected infrastructure runs from edge connected IoT devices to the cloud, sharing information across individual assets, plants, enterprises, and communities of connected services.

However, the final transformation does not have to be the first implementation. Like any solution, digital tools evolve as our needs and understanding of the problem change. Our use of the data infrastructure will necessarily evolve by:

- Collecting data into the data historian
- Contextualizing an operational "chart of accounts"
- Integrating with external data sources such as LIMS, ERP, CMMS, etc.
- Formulating streaming real-time analytics, first law calculations, and KPI's
- Detecting and notifying on errors and events
- Leveraging complex analytic engines to develop predictive and prescriptive models

Data should be collected once, as close to the source as possible. The data infrastructure will interface with these source systems. As part of the transformation effort, organizations evaluate missing data for the potential value of the information against the expense of collection. Where possible, organizations can leverage IoT technologies or collect manually.

Demonstrated use cases include monitoring site heat index to prevent heat stress, extrapolating corrosion curves to facilitate mechanical integrity programs, performing both golden batch and multivariate analyses, and collecting asset utilization data for reporting among others. Successful digital transformations provide competitive advantages for these enterprises. We believe people with data can transform our world.

590961 How to Realize Your Digital Transformation

Abstract:

Today, few companies doubt the benefits of Industry 4.0. It is relatively easy to talk about the need for a digital transformation from a high level, but when it comes to planning and executing a specific digital transformation roadmap, it tends to be more challenging. But it doesn't have to be that way. Industry 4.0 has now been around for a few years, which has allowed tried and tested approaches and methodologies to emerge that can be applied to pretty much any size company in any industry.

We will look at the value chain of chemical processing plants identifying application areas for Industry 4.0 technologies such as digital twins. The focus of this industry specific approach is on structured and open data models that break down stand-alone and isolated knowledge silos - in effect integrating the industry relevant value chains. Actionable items include, but is not limited to: the use of multiple digital twins as controlled copies handling business critical information for data-based business management; the continuous assessment of the correct technologies and technology levels to be implemented for each business relevant area; business and site level consulting on industry 4.0 until maturity is reached and digitalization is business as usual in the chemical processing industries

In addition, we will gain insights into how to shape your own future by combining novel and mature technologies in new and exciting ways, which includes project prioritization. It is a critical exercise for companies in formulating a digital transformation because it enables them to identify areas that will generate the greatest value and evaluate each of them based on specific criteria. These can then be prioritized, which will help decision-making and effective resource allocation. This may sound like a simple step, but it requires dialog, involvement from several stakeholders and a framework to put it all together. Industry 4.0 is *not* a one-size-fits-all approach.

By attending this presentation, you will gain the following insights:

- Gain high-level insight into the concept of 4.0 technologies applied in chemical processing industries.
- Gain site-level insight into digitalization use cases transforming our industry now!
- Basic understanding of how to identify Industry 4.0 projects.

View of how to best prioritize them in terms of resources, ROI, digital maturity level and more.

589365 Piloting Digitalization at Dupont

Abstract:

Okay, you've decided that it's critically important to begin going down the digitalization path. Now what? This presentation will cover DuPont's efforts to date to implement pilot digitalization projects making use of data from existing assets and provide some learnings that you may benefit from in your own piloting efforts.

Specific questions that will be answered and topics that will be covered include...

- Implementation in a production facility making use of real world data
- Selection of digitalization options to be piloted
- Who was involved in the implementation process?
- How long did it take to complete the piloting?
- What support did you receive during the piloting?
- What specific digitalization tools/apps were piloted?
- What was the outcome of the pilot testing?
- What would DuPont have done differently?
- What are the next steps?

591209 How to Network

Abstract:

Wanna land your dream job? Get ready to network! Most jobs and other game-changing career opportunities are not advertised, and even if they are, there is usually a short-list of candidates already in mind. So how do you find out about and access the 90% of jobs and other opportunities that are "hidden"? In this presentation, we will focus on proven networking strategies and tactics to identify new opportunities, locate decision-makers within organizations, solidify your reputation and brand in the minds of those who hire, and gain access to hidden jobs and game-changing opportunities. Discover how networking and self-promotion can enable you to land or even create your dream job from scratch!

590251 Panel Discussions - Future of Olefins

Abstract:

Olefins industry has been historically used to cyclic nature of business following the economic cycles. The regional growth has been driven by the regions with demand side or by advantaged feedstock. There are many historical and unique factors at play that will influence the future of olefins industry. In this session, we have invited industry leaders to share their ideas and thoughts about the factors that can significantly impact the industry in mid to long term. Some of the factors that may drive the future include:

- Historical factors like GDP growth, population growth, urbanization trends, quality of life improvements etc.
- Geo-political factors like trade barriers, tensions in Middle East, sanctions etc.
- Environmental pressures like recycling/recovery of plastics, carbon taxation/greenhouse gas emissions, ban of single use plastics etc.
- Changing energy dynamics driven by focus of oil majors away from fuels to petrochemicals
- Large and complex facilities with high capital intensity resulting in challenges to develop, build and operate them safely and achieving desired performance and financial outcomes
- Disruptive technologies like shale developments, transportation systems, sharing platforms, alternate energy sources etc.

590298 Distillation Column Control

Abstract:

The understanding of the basic control schemes on a distillation column is key for effective operation and troubleshooting. This paper introduces distillation column control, and the concepts behind typical control schemes employed in industrial distillation columns. The advantages, disadvantages, and pitfalls of control configurations will be illustrated using examples.

590409 Radioisotope Techniques for Process Investigations

Abstract:

In all chemical plants and refineries there generally is not enough instrumentation in the right places in
our processes. During the design phase, decisions were made between knowing more about the process,
which means buying more instruments, and saving capital funds. When a problem appears with the operation of a vessel, there are always gaps in the knowledge of what is happening due to a lack of instruments in the places where knowledge is not routinely required. Process engineers and operations personnel need to know what is happening so they can take corrective action. When knowledge is missing, they often take actions that make matters worse.

For distillation and separation processes Chemical Engineers have historically relied upon plant process
measurements such as flow rates, temperatures, and pressures and model or simulation results for data with which to solve troubleshooting or re-design problems. An array of on-line diagnostic services are available that can provide additional data offering real-time information on how pieces of process equipment are actually operating. The most common application is Gamma Scanning. This test is primarily applied, but not necessarily limited to, distillation or separation columns. Gamma scans provide a density profile of the internal process of operating distillation columns and other process vessels. The density profile can be used to diagnose the hydraulic operating conditions of mass transfer
devices such as damage to internals, flooding, degree of entrainment or weeping, liquid levels on trays and distributors, liquid distribution through packed beds, etc. The presenter will show case studies where scanning revealed vital process information that helped solve an operating problem, or helped make a revamp/re-design successful. Additional derivatives from gamma scanning of detecting flow distribution patterns will be demonstrated. The ThruVision Scan technique is a specialized horizontal gamma scan used to generate a topographic profile of the internal cross-sectional density of process equipment. This profile is useful for the detailed study of liquid flow distribution through packed columns as well as having applications beyond distillation columns. A hybrid application involving tracers
and scanning on fixed bed catalytic reactors will also be discussed.

591211 How to Reduce Conflict to Keep Your Job

Abstract:

Conflict is a natural part of team engagement, but there are specific ways to manage, mitigate, and possibly even prevent conflicts from arising. In this presentation, we will discuss how to handle conflicts and forge a relationship with the other party that demonstrates your professionalism, and your interest in collaboration. Communication is the key, of course, and case studies will be analyzed.

591159 When Disasters Happen: A Collaborative Approach Towards the Community's Recovery Process

Abstract:

The 2020 Virtual AIChE Spring Meeting & 16th GCPS Wednesday keynote "When Disasters Happen: A Collaborative Approach Towards the Community's Recovery Process" will be presented by Thomas Munoz, Emergency Manager and Homeland Security Director with Texas City. As the Emergency Manager and Homeland Security Director with Texas City, the home of 19 chemical plants, he is responsible for the safety of all its citizens along with managing the crisis communication plan with local industry and major stakeholders.

Safety and disaster planning is vital to the recovery process after a major incident. As key critical infrastructure industries, issues may arise that have not been planned for in advance, such as the recovery from an incident. Developing and maintaining partnerships with local partners in your community will support addressing unforeseen challenges. Recovery efforts include planning with key partners and understanding the phases of recovery, which also consider the mental health and well of everyone involved in the response, as well as the community impacted. In this keynote address a discussion of pre-incident planning considerations and strategies as well as lessons learned/after action improvements post-recovery will be discussed. Additionally, collaboration considerations regarding local resources to promote recovery in the event of a workplace incident will be discussed.

C0-presenter Amanda Groller, MS.Ed., LPC will discuss the mental health impacts of a disaster.

589938 Hydrogen As a Vector for the Energy Transition

Abstract:

Hydrogen can play a key role as an energy vector for difficult-to-decarbonize industries and sectors including heavy duty trucking and transport, marine shipping, aviation, and industry heat and power. Integrating usage in these sectors is key to obtaining economies of scale needed for commercial roll out, via sharing of infrastructure. Process safety with global consistency in codes and standards for infrastructure and use will be critical. This presentation will address the role hydrogen will play in synergy with electrification as a vector for end-user delivery of clean energy options in the future economy, and the importance of process safety in gaining broad stakeholder acceptance needed for adoption.

591124 Starting and Maintaining a Business

Abstract:

In this presentation I will share my experience in starting and successfully operating a consulting business in environmental, health and safety. I started a consulting business ten years ago and since have managed to successfully generate increasing revenue stream while maintaining good profit margins.

This talk will discuss how to carve a niche which has a demand in service business, trials and tribulations in starting an enterprise from scratch including generating initial revenue, obtaining appropriate insurance and resources required both financial and human. I will also discuss my experience in setting up company name and logo.

Maintaining and sustaining a small service business requires being both a" seller" and "doer". I will discuss balancing between need to sell work as well as complete technical projects. I will discuss my own overall approach to generating steady revenue stream. I will relate my experience in maintaining cash flow, balance sheet and other financial issues.

I will discuss some of the attitude and aptitude required to be successful in business based on my experience. There is a lot of satisfaction and a sense of accomplishment in operating a business if there is a methodical approach with right attitude and ability to adapt to changes.

588706 Starting a Consulting Business: Lessons Learned from 18 Years of Being *Self-Unemployed*

Abstract:

The "traditional employment model" is shifting: retiring "Baby Boomers" are departing the Hydrocarbon Processing Industry (HPI) in droves, taking decades, even centuries, of hard-earned knowledge and expertise with them. On the other end of the spectrum, short-sighted management policies, promulgated over the preceding twenty-plus years, have left a large age and experience gap that separates the in-coming "Millennials" from the Boomers. That is, there has been grossly insufficient recruitment, training & mentoring and retention to adequately replace retirees and to train and mentor the next generation. This situation presents a unique opportunity for the engineer turned entrepreneur to start a consulting business.

While the opportunity for becoming an independent consultant and having a successful business exists in the marketplace, the journey to success is fraught with numerous traps, distractions and pitfalls for the unprepared and unaware. For the curious, as well as the serious explorer, this presentation is a tale of the trials, tribulations, successes and observations of an 18-year independent consultant veteran and will address the skills, strategies and dedication needed to be successfully "self-unemployed".

591219 Turning a Calling into an Award Winning Business

Abstract:

Strata-G, LLC was founded in 2002 with a vision for environmental stewardship and a calling to be a business that honors the simple ideal of treating people the way we want to be treated. By staying committed to this ideal, Strata-G has grown into a successful professional services team of approximately 240 engineers, scientists, safety professionals, and project managers with a culture that is often described as family. Strata-G's dedication to its core values of integrity, service, and quality has resulted in more than 30 company awards from clients and the community for performance and service, including five Veteran-owned Business of the Year Awards, the Innovation Valley Navigator Award for Community Impact, the 2019 Small Business Administration Region IV Subcontractor of the Year Award, and the 2019 US Chamber of Commerce Dream Big Award as National Small Business of the Year. Dan Hurst, Founder and CEO, will discuss launching a bootstrap startup, building a team, establishing a culture, overcoming fears and failures, and identifying success principles which somehow resulted in a thriving business recognized nationally for its community impact, employee engagement, and customer service.

592043 Unconscious Bias: Fostering Inclusive Communities

Abstract:

Part I.
WELCOME & INTRODUCTION
Facilitated by: Cynthia Murphy-Ortega, Manager of University Partnerships and Association Relations
1. Engineering current state vs. next generation
2. Define key Diversity, Equity and Inclusion terms
   What is Diversity vs. Inclusion and What does Equity look like?
   Define Unconscious Bias
3. Brief Q&A
4. Introduce panel moderator
Part II.
Panel Discussion, Engaging Chemical Engineering Communities to achieve Equity, Diversity, and Inclusion
Facilitated by: Dina Clark, Head of Diversity & Inclusion, North America, Covestro
A practical discussion on equity, diversity, and inclusion in engineering; featuring experiences from senior leaders of the engineering community
▪ Garry Bryan, Product Supply Principal Expert at Bayer Crop Science
▪ Deborah L. Grubbe, PE, CEng., NAC, 2020 President-Elect – The American Institute of Chemical Engineers (AIChE)
▪ Belinda Akpa, Assistant Professor of Integrated Synthetic and Systems Biology, NCSU
Part III.
Question & Answer
Program Concludes

589696 Analysis of Hydrogen Cylinder Explosion Accidents and Safety Disposition of Remaining Cylinders

Abstract:

Hydrogen is widely used in process industry because of its various application. Cylinders are important instruments to carry hydrogen. However, accidents of hydrogen cylinders often occur during the hydrogen production, operation, transportation, and storage. On July 23, 2004, three quartz manufacturing companies' hydrogen cylinders exploded during operation in the same county of China. Six people were killed and several buildings collapsed. In this paper, the cause of hydrogen cylinder explosions is analyzed by introducing the production and operation process of hydrogen. Chemical explosion is determined from the cylinder debris and the chemical composition of gas in the remaining cylinders. The energy and the overpressure generated by the explosion are estimated through calculation. Personnel injury and building damage are also analyzed. In order to safely release hydrogen, corresponding method and devices are designed. The lessons learned from chemical process safety are also presented to prevent similar accidents from occurring.

Tubular high temperature PEM fuel cell with novel 3D-printed metallic porous anode Gas Diffusion Layer: fabrication, model development and validation

María Catalina Bermúdez Agudelo, Aaron D. Ratschow, Manfred J. Hampe,
Thermal Process Engineering Group, Technical University of Darmstadt, Darmstadt, Germany

Abstract

This paper presents a fabrication method of single concentric multilayer tubular high temperature PEM fuel cell using a novel 3D-printed metallic porous anode gas diffusion layer (GDLa) produced via selective laser melting (SLM) process. Furthermore, a two-dimensional axisymmetric non-isothermal steady-state model is developed. This model considers the major transport phenomena as well as the fuel cell loss mechanisms, while the mixed potential is neglected. The membrane electrode assembly (MEA) was modeled analytically using a thin-layer approach and the Butler-Volmer equation to describe both electrochemical reactions. The model was implemented in the commercial software COMSOL Multiphysics® and validated versus experimental data. Additionally, simulation results were compared with an available detailed model. There was a good agreement with both experimental data and results from the detailed model, demonstrating the applicability of the developed model to predict the complex fuel cell internal processes and fuel cell performance of the fabricated fuel cell.

Nomenclature

A	area	M	molar mass
a	surface area per catalyst unit mass	n	number of transferred electrons
C_p	heat capacity	p	pressure
D	diffusivity	$Pt_{loading}$	platinum mass per unit area
d	diameter	Q	heat source term
S	reaction entropy	R	ideal gas constant
E	electric potential	r	radius
F	Faraday's constant	R_j	electric resistance
j	current density	R_{total}	total electric resistance
j_0	exchange current density	S_m	mass source term
K	roughness factor	T	temperature
k	thermal conductivity	u	velocity
k_{int}	intrinsic reaction rate	V	volume
k_p	permeability	V_{cell}	cell voltage
L	length	x	mole fraction
Greek letters			
α	transfer coefficient	ϵ	porosity
γ	pressure coefficient	η	overpotential
μ	viscosity	ξ	catalyst effectivity

ρ	density	ϕ	volume fraction
σ	electrical conductivity	ω	mass fraction
Φ	Thiele modulus		

Superscripts

0	standard	ref	reference
eff	effective		

Subscripts

act	activation	H_2O	water
An	Anode	H_3PO_4	phosphoric acid
Ca	cathode	i	index
Cat	catalyst	in	inner
CL	catalyst layer	j	index
$diff$	diffusion	o	outer
ele	electrolyte	O_2	oxygen
eq	equilibrium	Pt	platinum
f	fluid	R	reaction, reactant
H_2	hydrogen	s	solid

Introduction

The high-temperature proton exchange membrane fuel cell (HT-PEM-FC) is considered a promising candidate for energy generation in stationary and non-stationary applications mostly due to its middle working temperature, compactness, and high power density. The plate-and-frame design is the most widely used cell configuration, in which a significant portion of the manufacturing cost falls on components such as bipolar plates, current collectors, and gas diffusion layers. To overcome this problem, researchers have introduced 3D-printing technology into the field for the production of bipolar plates and gas diffusion layers to some small extent. Among the 3D-printing methods, the selective laser melting (SLM) process permits the production of complex structures, such as those required in fuel cell technology. Although the SLM process is often used to create dense objects, a variation of process parameters leads to porous elements [1–3]. A further cost reduction is expected if alternative fuel cell configurations are considered. For instance, the well-known tubular design commonly used in the solid oxide fuel cells, in which case the bipolar plates can be omitted.

In recent years, researchers have investigated tubular designs of low temperature [4–12] and high temperature [13] PEM fuel cell systems both experimentally and virtually. This work presents a fabrication method and a 2D simulation model to describe single tubular HT-PEM-FCs. The method is in line with the procedure used by Lang [13] with some modifications during fabrication and assembly. The developed model considers the major transport phenomena as well as the fuel cell loss mechanisms, namely activation losses, ohmic losses, and mass transport losses, while the mixed potential is neglected. To model the membrane electrode assembly (MEA) a novel thin-layer approach is used, in which the Butler-Volmer equation is used to describe both electrochemical reactions. The model was validated versus experimental data and was compared with an available detailed model for temperature, anodic, and cathodic gas inflow. The goal of this work is to confirm the suitability of

SLM as an additive manufacturing process for GDL construction and confirm the validity of a simplified model for fuel cell investigation.

Experimental

Fuel Cell Fabrication Method

In cooperation with the Institute of Production Management, Technology and Machine Tools (PTW) a 3D-printed metallic porous GDL was produced via a Selective Laser Melting (SLM) process using the stainless steel 316L. The resulting thin-walled GDL is used in the assembly also as a hydrogen channel and current collector in the anode side. It was designed in such a way that a defined porosity could be achieved with an active area of 4.78 cm^2. Lang [13] used in his investigation similar electrodes with some differences in material, process parameters, wall thickness and porosity.

Catalyst inks were prepared by ultrasonication of a commercial catalyst powder (HiSPEC® 3000), a Polytetrafluoroethylene (PTFE) 50wt.% solution (Dyneon™ TF 5070GZ), and a mixture of isopropyl alcohol/water for at least 30 minutes. The catalyst layers were coated by the spray coating technique. The anode catalyst layer was sprayed onto the metal GDL, while the cathode catalyst layer was sprayed later onto the surface of the high-temperature membrane.

The used membrane corresponds to a meta-Polybenzimidazole (m-PBI) membrane doped with phosphoric acid (H_3PO_4). To achieve higher doping levels, the polyphosphoric acid (PPA) method was employed following the sol-gel procedure described by Xiao et al.[14] and Lang [13]. Once a satisfactory solution was achieved, the membrane was applied to the anode catalyst layer via a dip-coating technique and hung for some hours at ambient conditions. The excess of acid was removed by rinsing the membrane with ca. 3 ml of deionized water.

The commercial woven carbon hose (KB-4002) was employed as cathode GDL. The multilayer construction was fixed using a lathe machine and a metal wire, such as tungsten. This step assures a uniform compression and an electric connection. To isolate the fuel cell components electrically, a PTFE tape was used.

Figure 1 illustrates the schematic procedure to fabricate the multilayer concentrically tubular arrangement used in this study. As can be seen, this assembly omitted typical fuel cell components, such as flow field channels, end plates, current collectors, and gaskets, which suggests a considerable reduction in time and cost of fuel cell manufacturing. The necessary compression and electrical conduction are provided with a wrapped conductive metal wire that covers the active area uniformly.

Fuel Cell Characterization

The fuel cell is electrochemically characterized using a commercial FuelCon Evaluator C100 test station through polarization curves and electrochemical impedance spectroscopy (EIS) with hydrogen quality 3.0 as fuel and compressed air as oxidizer at atmospheric pressure (1 bar). The polarization curve was recorded in galvanostatic and potentiostatic modes in a voltage range from the open-circuit voltage (OCV) down to 0.24 V. The EIS measurements were performed at a constant load of 0.1 A in a frequency range from 0.1 Hz to 20 kHz. The obtained fuel cell resistances were used in the simulation model as parameters.

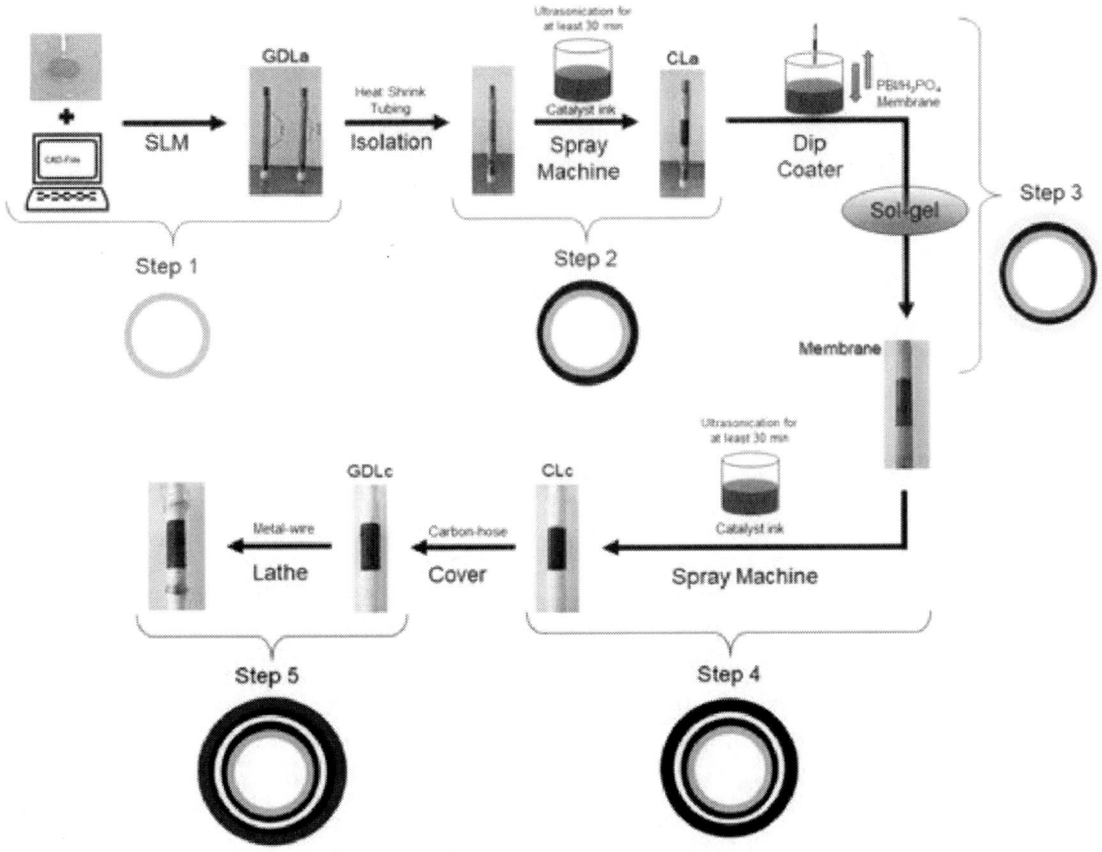

Figure 1. Fabrication method for multilayer concentric tubular HT-PEM-FC

Developed Fuel Cell Model

Model Descriptions and Assumptions

Figure 2 illustrates the computational domains of the developed 2D axial-symmetrical model of a single tubular HT-PEM-FC. The model geometry includes the tubular reactor and thermal isolation used for fuel cell characterization. The MEA is analytically modeled using a thin layer approach. The developed model considers the following assumptions: (i) steady-state, (ii) ideal gas mixture, (iii) non-volatile phosphoric acid, (iv) no mixed potential, (v) solid materials are homogeneous and isotropic, (vi) water is gaseous.

Governing equations

All governing equations and source terms considered in this model are listed in Table 1. The model is solved for the velocity by the combination of the continuity equation with the Navier-Stokes and the Brinkman equations in all porous domains, i.e. anode GDL and cathode GDL. The concentration of species is modeled using the Stefan-Maxwell diffusion model with the Bruggeman correction for diffusion coefficients. The energy balance is solved in all domains. In solids, the energy transport is modeled as heat conduction, while in gas channels and porous domains a convective term is added. The transport equations are coupled to account changes in velocity, density, and temperature. The thin layer approach allows all source terms associated with the electrochemical reactions to be considered as boundary fluxes and mathematically described as Neumann boundary conditions.

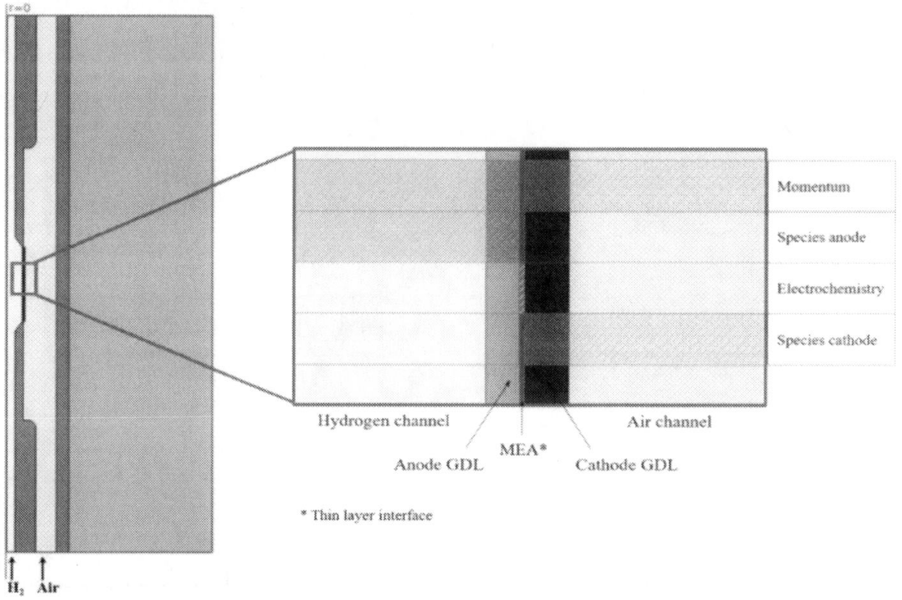

Figure 2. Model geometry and mathematical equations not including the energy equation.

MEA modeling: Thin layer approach

The Butler-Volmer equations for anode (Equation 13) and cathode (Equation 14) are combined with the operating voltage equation (Equation 15) to solve analytically the thin layer interface used to represent the MEA in this study. This combination leads to a nonlinear system with three equations and four unknown variables, namely: j_{net}, V_{Cell}, $\eta_{act,An}$ and $\eta_{act,Ca}$. This system is coupled with the computational fluid dynamics (CFD) model and solved by sweeping the cell voltage.

$$j_{An} = j_{net} = j_{0,An}\, \eta_{Cat,An}(j_{net})\, K_{Cat,An}\left(e^{\alpha_{An}\frac{nF}{RT}\eta_{act,An}} - e^{-(1-\alpha_{An})\frac{nF}{RT}\eta_{act,An}}\right) \quad (13)$$

$$j_{Ca} = j_{net}\, A_{CL,An}/A_{CL,Ca} = j_{0,Ca}\, \eta_{Cat,Ca}(j_{net})\, K_{Cat,Ca}\left(e^{\alpha_{Ca}\frac{nF}{RT}\eta_{act,Ca}} - e^{-(1-\alpha_{Ca})\frac{nF}{RT}\eta_{act,Ca}}\right) \quad (14)$$

$$V_{Cell} = E_{eq} - \eta_{act,An} - \eta_{act,Ca} - \eta_{ohmic}(j_{net}) \quad (15)$$

The aforementioned variables are defined and tabulated in Table 2.

Table 1. Governing equations and source terms

Governing equations			Source terms	
Energy	Heat transfer equation $\nabla \cdot (-k\, \nabla T_s) = 0$	(1)	$Q_{R,An} = \underbrace{j_{An}\left(\eta_{act,An} + \frac{T\, \Delta S_{R,An}}{2F}\right)}_{\text{Generation by HOR}}$	(7)
	$\nabla \cdot (-k\, \nabla T_f) = -\rho\, C_p\, \vec{u} \cdot \nabla T_f$	(2)	$Q_{R,Ca} = \underbrace{j_{Ca}\left(\eta_{act,Ca} + \frac{T\, \Delta S_{R,Ca}}{4F}\right)}_{\text{Generation by ORR}}$	(8)
			$Q_{Joule} = \underbrace{j_{net}^2\, A_{CL}\, R_{total}}_{\text{Joule heating}}$	(9)

Momentum	Continuity equation $\nabla\vec{u} = 0$	(3)	$S_{m,H_2} = \underbrace{-\frac{j_{An}\, M_{H_2}}{2F}}_{\text{Hydrogen sink}}$	(10)
	Navier-Stokes equation $\rho(\vec{u}\cdot\nabla)\cdot\vec{u} = \nabla\cdot\left(-p\cdot I + \eta\cdot(\nabla\vec{u} + (\nabla\vec{u})^T)\right)$	(4)	$S_{m,O_2} = \underbrace{-\frac{j_{Ca}\, M_{O_2}}{4F}}_{\text{Oxygen sink}}$	(11)
	Brinkman equation $\left(\frac{\eta}{k_p}\right)\vec{u} = \nabla\cdot\left(-p\cdot I + \frac{1}{\epsilon}\left(\eta(\nabla\vec{u} + (\nabla\vec{u})^T) - \left(\frac{2}{3}\eta\right)(\nabla\cdot\vec{u})\cdot I\right)\right)$	(5)		
Chemical species	Stefan-Maxwell equation with Bruggeman correction $\nabla\cdot\left(\rho\,\omega_i\,\vec{u} - \rho\,\omega_i\sum_{j=1}^{n}\tilde{D}_{ij}^{eff}\left(\nabla x_j + (x_j - \omega_j)\frac{\nabla p}{p}\right)\right) = 0$	(6)	$S_{m,H_2O} = \underbrace{\frac{j_{Ca}\, M_{H_2O}}{2F}}_{\text{Water source}}$	(12)

Table 2. Variables used for the thin layer approach

Exchange current density	$j_0 = j_0^{ref}\left(\frac{p_R}{p_R^{ref}}\right)^{\gamma} exp\left(-\frac{E_{act}}{RT}\left(1 - \frac{T}{T_{ref}}\right)\right)$		(16)
Effectiveness factor	$\eta_{Cat} = \frac{3}{\phi}\left(\frac{1}{\tanh\phi} - \frac{1}{\phi}\right)$	→ To quantify the concentration losses	(17)
Thiele modulus	$\phi = L\sqrt{k_{int}/D_{ij}^{eff}}$		(18)
Intrinsic reaction rate	$k_{int} = \frac{j_{net}}{nF}\frac{A_{CL}}{V_{CL}}\cdot 1\left[\frac{m^3}{mol}\right] = \frac{j_{vol}}{nF}\cdot 1\left[\frac{m^3}{mol}\right]$		(19)
Roughness factor	$K_{Cat} = Pt_{loading}\; a_{Cat}$		(20)
Specific electrochemical surface area	$a_{Cat} = \frac{6}{d_{Pt}\,\rho_{Pt}}$		(21)
Equilibrium Potential	$E_{eq} = E^0 + \frac{\Delta S}{nF}(T - T^0) - \frac{RT}{nF}\ln\frac{(p_{H_2O}/p^0)}{(p_{H_2}/p^0)\;(p_{O_2}/p^0)^{0.5}}$		(22)
Ohmic losses	$\eta_{Ohmic} = R_{total}\, j_{net}\, A_{CL}$		(23)
Total resistance	$R_{total} = \sum_{ionic,electric} R_j$		(24)
Resistances in hollow cylinders	$R_j = \frac{\ln(r_{o,j}/r_{in,j})}{\sigma_j^{eff}\; 2\pi L_{Cat}}$		(25)

Brugemann corrections	$D_{ij}^{eff} = D_{ij}\epsilon^{1.5}$ $\sigma_j^{eff} = \sigma_j\,(1-\epsilon)^{1.5}$ → corrected using porosity $\sigma_{H_3PO_4}^{eff} = \sigma_{H_3PO_4}\,\phi_{ele}^{1.5}$ → corrected using electrolyte volume fraction	(26)
Empirical ionic conductivity [13]	$\sigma_{H_3PO_4} = -0.091 - 0.631\omega_{H_3PO_4} + 0.000246T - 0.159\omega_{H_3PO_4}^2$ $+0.00464\omega_{H_3PO_4}T - 1.647\omega_{H_3PO_4}^3 + 0.00213\omega_{H_3PO_4}^2 T$ $+1.411\omega_{H_3PO_4}^4 - 0.00325\omega_{H_3PO_4}^3 T$	(27)

Results and Discussion

Following the proposed fabrication method, functional fuel cells were manufactured and characterized. The results showed that SLM is a feasible alternative for GDL production. The implemented changes in materials, material properties, and process parameters had a positive impact on fuel cell performance. However, further investigations are needed to accelerate the commercialization of this type of fuel cell. The developed model is simulated using the finite element commercial software COMSOL Multiphysics® version 5.4. A total of 46.921 degrees of freedom was solved using the MUMPS solver in one single study step. To validate the model, the simulation data was compared with both the results obtained from the detailed model proposed by Lang [13] and the corresponding experimental data. The experiments were performed according to the aforementioned procedure using an aged fuel cell with a platinum loading of 0.58 mg/cm^2 and 0.54 mg/cm^2 for anode and cathode side, respectively. The predicted polarization curves take into account the degradation associated with an aged fuel cell in parameters such as fuel cell resistances. However, the effect of fuel cell degradation on the OCV is not displayed, since the mixed potential has been neglected in both the present model and the detailed model of Lang [13].

Table 3. RSME and R^2 for different operating parameters.

	Developed model		**Detailed model**	
	R^2	**RSME [A/cm^2]**	**R^2**	**RSME [A/cm^2]**
Figure 3 (a)	0.993	0.00220	0.994	0.00472
Figure 3 (b)	0.997	0.00248	0.996	0.00762
Figure 3 (c)	0.999	0.00286	0.999	0.00584

Figure 3 presents the comparison between the predicted polarization curves with the measured values for different operating parameters. A good agreement between model curves and experimental data is observed. The deviations at the OCV between experimental data and simulation results suggest the presence of cross-over losses during characterization. Also, the comparison shows slight differences in the high current density region for the detailed model with a maximum deviation of 67mV, which corresponds to a relative deviation of 19%. The mean square deviation (RSME or RSMD), as well as the coefficient of determination (R^2), are summarized in Table 3.

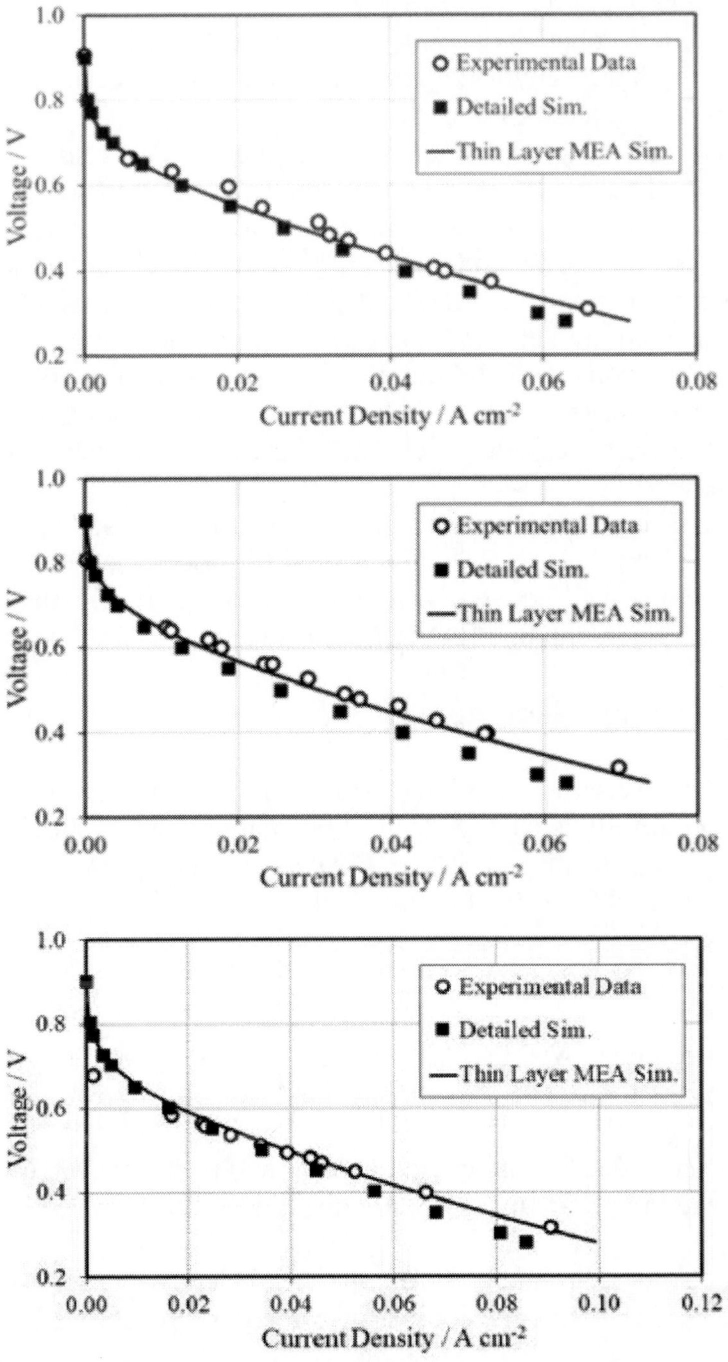

(a) T_{cell}= 120°C
Hydrogen inflow= 0.1 NLPM
Air inflow= 2.0 NLPM

(b) T_{cell}= 160°C
Hydrogen inflow= 0.2 NLPM
Air inflow= 2.0 NLPM

(c) T_{cell}= 160°C
Hydrogen inflow= 0.1 NLPM
Air inflow= 0.5 NLPM

Figure 3. Comparison between the developed model, detailed model, and experimental data at different operating parameters: (a) case study 1, (b) case study 2, and (c) case study 3.

Conclusion

In this work, a fabrication method for tubular HT-PEM FC using 3D-printed metallic porous anode GDL was presented. The application of 3D-printing technology in GDL production introduces an economical and fast manufacturing process that allows alternative fuel cell designs to penetrate the PEM fuel cell market. Following this procedure, a functional fuel cell was manufactured and electrochemically characterized at different operating parameters.

The developed model is two-dimensional, axisymmetric, non-isothermal, steady-state, and single-phase. It combined a CFD model with an analytical approach for the MEA based on a thin-layer interface. The operating voltage equation integrates the thermodynamics with the principal loss mechanisms, without taking into account the mixed potential, and uses the Butler-Volmer equation to describe both electrochemical reactions. The model was implemented in the commercial software COMSOL Multiphysics® and solved for velocity, pressure, mass fractions, temperature as well as local current densities and fuel cell losses (i.e. activation-concentration losses and ohmic losses). To assure a reliable fuel cell performance prediction, the simulated polarization curves were compared with the results obtained from the detailed model developed by Lang [13] and the corresponding experimental data. In both cases, the simulation data exhibited a good fit, despite slight differences in some cases at the OCV and lower density current region, indicating that mixed potential would be required for a better adjustment, especially in degraded fuel cells. Research into incorporating this effect in the model is planned for future work.

Some conclusion can be drawn based on the present findings. First, the proposed tubular design might be an appealing alternative to the planar design. Secondly, the SLM process can be an option to fabricate GDL for fuel cells in an easy and cost-effective manufacturing process. Lastly, the simplified developed model represents a helpful tool for evaluating fuel cell performance.

Acknowledgments

The financial support from the German Research Foundation (DFG) within the framework of projects AB 133/97-1 and HA 1283/11-1 is gratefully acknowledged. The authors thank the Institute of Production Management, Technology and Machine Tools (PTW) at the Technical University of Darmstadt for the production of the 3D-printed GDL used in this work.

References

1. Abele E, Reiber T, Hampe M, Bermúdez Agudelo MC, Menz F. Tailoring and investigation of defined porosity properties in thin-walled 316L structures using laser-based powder bed fusion. Progress in Additive Manufacturing. 2019;4(4):451–463.
2. Peng T, Chen C. Influence of energy density on energy demand and porosity of 316L stainless steel fabricated by selective laser melting. Int. J. of Precis. Eng. and Manuf.-Green Tech. 2018;5(1):55–62.
3. Abele E, Stoffregen HA, Kniepkamp M, Lang S, Hampe M. Selective Laser Melting for Manufacturing of Thin-walled Porous Elements. Journal of Materials Processing Technology. 2014;(215):114–121.
4. Lee KI, Chung DK, Park MS, Chu CN. The development of a cylindrical proton exchange membrane fuel cell with an integrated metal-hydride container. Int. J. Precis. Eng. Manuf. 2013;14(6):1065–1070.

5. Modekurti S, Bullecks B, Bhattacharyya D, Rengaswamy R. Modeling Studies of a Cylindrical Polymer Electrolyte Membrane Fuel Cell Cathode. Ind. Eng. Chem. Res. 2012;51(13):5003–5010.
6. Bullecks B, Rengaswamy R, Bhattacharyya D, Campbell G. Development of a cylindrical PEM fuel cell. International Journal of Hydrogen Energy. 2011;36(1):713–719.
7. Lee KI, Lee SW, Park MS, Chu CN. The development of air-breathing proton exchange membrane fuel cell (PEMFC) with a cylindrical configuration. International Journal of Hydrogen Energy. 2010;35(21):11844–11854.
8. Sadiq Al-Baghdadi MAR. Three-dimensional computational fluid dynamics model of a tubular-shaped PEM fuel cell. Renewable Energy. 2008;33(6):1334–1345.
9. Evans RB. Modeling and Testing of a Micro-Tubular Low- Temperature Fuel Cell for use in a Micro Air Vehicle. Dissertation. Virginia: Virginia Tech, 2007.
10. Yazici MS. Passive air management for cylindrical cartridge fuel cells. Journal of Power Sources. 2007;166(1):137–142.
11. Coursange J-F, Hourri A, Hamelin J. Performance Comparison Between Planar and Tubular-Shaped PEM Fuel Cells by Three-Dimensional Numerical Simulation. Fuel Cells. 2003;3(12):28–36.
12. Green KJ, Slee R, Lakeman JB. The development of lightweight, ambient air breathing, tubular PEM fuel cell. Journal of New Materials for Electrochemical Systems. 2002;5:1–7.
13. Lang S. Entwicklung tubularer Mitteltemperatur-Brennstoffzellen - Experimentelle Untersuchungen, Modellierung und numerische Simulation. Verfahrenstechnik. München: Dr. Hut, 2016.
14. Xiao Lixiang, Zhang Haifeng, Scanlon Eugene, Ramanathan L.S., Choe Eui-Won, Rogers Diana, Apple Tom, Benicewicz Brian C. High-Temperature Polybenzimidazole Fuel Cell Membranes via a Sol-Gel Process. Chemistry of Materials. 2005;17(21): 5328–5333.

586150 A Novel Approach for Design and Quantitative Risk Assessment for Hydrogen Refueling Station

Abstract:

In order to move towards further commercialization of the Hydrogen economy, it is essential to minimize the associated safety concerns and maximize the profitability of the involved systems such as that of Hydrogen refueling stations. Currently, Hydrogen refueling stations are primarily designed to satisfy required consumer demand while ensuring that Hydrogen is supplied at a profitable rate. The developed design is then supported by a quantitative risk analysis (QRA) carried as per the NFPA-2 standard to validate that the associated risks (such as that of fire and explosion) are acceptable. However, the absence of QRA in the design stage itself leads to a possibility of the refueling stations being over-designed from the perspective of safety. As a part of this study, we have developed a novel risk-based design approach for Hydrogen refueling stations that allows the design of these facilities with optimized economics and risk. The Hydrogen refueling station is designed with a simultaneous QRA and economic analysis. However, The QRA methodology is modified to fit the needs of a Hydrogen refueling station. The population and the time spent by an individual (which are important parameters for an effective QRA) in the refueling station are considered as variables (unlike traditional QRA for chemical facilities that assumes these values as constants) using the queuing theory. The QRA methodology accommodates the hazards of explosion, jet-fire and asphyxiation for the employees as well as the customers present at the refueling station.

The results obtained from this study provide deeper insights into the risk-benefit counterbalance associated with the design of Hydrogen refueling stations and provides design guidelines to facilitate further commercialization of the Hydrogen economy.

592854 Mindful Methods to Create Profound Shifts in Personal and Corporate Life

Abstract:

In this session, Lisa M. Liszcz, Ph.D. will help you plan for more energy in your life! She will describe how the workplace is evolving in the #Metoo and #TimesUp era and discuss the need for balance between masculine and feminine energies within individuals, teams, and organizations.

Do you go home from work drained and exhausted? Is your mantra "I gave at the office"? Participate in this session, and you will take a brief proprietary quiz to find your unique *Energy Style*. You will make a plan to stay in your *Energy Style*. Never go home exhausted again.

Based on her best-selling book, *Feminine Emergence*™, Dr. Liszcz will outline practical steps for bringing masculine and feminine energies to life at work. Her insightful stories from her personal history, her interviews with forty-four professional mid-career women, and references to pop culture will show the value of bringing more feminine energy into our companies and organizations.

In her book, *Feminine Emergence,* Dr. Liszcz outlines what is missing from a life, a team, or an organization that is out of balance. She will give an overview of her book in this instructive presentation. This new and emerging view of life and work will inspire you to engage talents and gifts that are available to all of us for more collaboration, inclusion and productivity.

Attendees will learn how to:

- Understand masculine and feminine energy
- Engage employees more fully in their work.
- Facilitate teams in a way that leverages more talents and skills of members.
- Enhance organizational performance through development of a balanced energy culture and corporate culture.

592114 Process and adsorbent for separating ethanol and associated oxygenates from a biofermentation system

Abstract:

Abstract for poster presentation or talk:

Abstract Title: Process and adsorbent for separating ethanol and associated oxygenates from a biofermentation system

Presenter Information: Deepak Sharma; Sr. Process Engineer, Engineering and Technology, Bayer AG

Patent number: US8658845B2

Patent Authors:

Deepak Sharma - Sr. Process Engineer, Bayer Ag, St. Louis; (Presenting Author)

Anil Oroskar – CTO, Orochem Technologies, Napervielle, IL;

David W. House – Scientist, Orochem Technologies, Napervielle, IL;

Alice M. Havill – Scientist, Lanzatech, New Zealand

The principles of process intensification if used on existing processes can lead to development and redesign of existing processes to become more sustainable. This presentation will discuss a process and an adsorbent for the separation of ethanol associated oxygenates from a dilute mixture of ethanol and associated oxygenates in water in the presence of organic compounds derived from a biofermentation process. After pretreatment, the separation is carried out in a simulated moving bed adsorption system employing a stationary phase adsorbent comprising fluorinated carbon or modified C18 silica gel selective for the adsorption of ethanol and associated oxygenates, such as 2,3-butanediol, with a mobile phase desorbent selected from the group consisting of methanol, ethanol, propanol, and methyl tertiary butyl ether. The process is useful for removing water from dilute aqueous mixtures of organic compounds comprising ethanol in dilute concentration in water and produced by fermentation, biomass extraction, biocatalytic, and enzymatic processes which are not economically recoverable by conventional distillation methods.

Keywords: Separation; Adsorption; SMB; Purification

Patent Assignee: Orochem Technologies LLC

592111 Pump safety, the undiscussed topic in chemical industry leading to several fatal incidents

Abstract:

Abstract Title: Pump safety, the undiscussed topic in chemical industry leading to several fatal incidents

Presenter Information: Deepak Sharma; Sr. Process Engineer, Engineering and Technology, Bayer AG

Pumps are one of the equipment's in the process industry that require a thorough review during HAZOP studies. The constant source of energy given to the pump, if not converted to kinetic energy to flow the fluid, can lead to heating of the pumped fluid. If the fluid keeps accumulating the energy provided by the motor, in a blocked-in scenario, it can lead to very high temperatures eventually leading to loss of containment of the pump and the pump can get damaged. There have been instances in the industry where fragments of damaged pumps have flown several feet away from the point of loss of containment leading to injuries. The poster will outline a scenario where a pump is run blocked in on suction and discharge for water and a chemical which has a low thermal decomposition temperature. Arc data is used to evaluate the onset temperature of thermal decomposition. The poster will outline safeguards that can be employed to avoid catastrophic loss of containment of the pump. Outlined below are safeguards used to avoid loss of containment of a pump in case of a blocked-in scenario:

- Install PSV on suction of pump
- Employ a low amp/power interlock on the pump to turn off the pump in case of blocked in situation
- Have a high-pressure interlock on the discharge line of the pump to direct flow back to the tank in case of pump deadhead
- Install an Interlock to turn off the pump in case the feed forward valves (automatic valves) close inadvertently

Pump deadhead scenarios are not only catastrophic when in flammable service there are several salts that have low thermal decomposition temperatures. Pumps in water service left blocked in could also blow catastrophically leading to reportable incidents. I feel this is an important topic to bring to the attention of students and professionals in the chemical industry. Poster will include pictures of actual installed pumps along with required safeguards in the field along with supporting P&ID depictions.

589440 New Kind of Linear Nanocomposite Applied in Improving Rheological Behavior of Crude Oil and Its Mechanism Studied through Molecular Simulation Method

Abstract:

In this work, a new kind of linear nanocomposite produced by carbon nanotube(CNT) as pour point depressant (C2-100) was developed by blending with MCNT nanoparticle and copolymer named Ethylene-vinyl acetate (EVA). Its effect of reducing yield stress and modifying the morphology of wax agglomerate were also evaluated. Besides, the synergetic effect of C2-100 nanocomposite with the magnetic field was also conducted. The mechanism of modifying rheological behavior of waxy crude oil was studied through molecular simulation method, which aims at understanding the interaction between this nanocomposite and wax molecule. The results indicated that C2-100 nanocomposite could reduce the yield stress of waxy crude oil more effectively than the copolymer (EVA) even at low concentration, as well as changing the morphology of wax crystals and agglomerate. Moreover, its effect of reducing yield stress becomes more obvious when the alternate magnetic field was applied. In addition, the results of molecular simulation also presented that the effect of improving rheological behavior of C2-100 nanocomposite is related to the adsorption behavior of wax molecules on the surface of MCNT nanoparticle, which is adsorbed with EVA molecules. In addition, the adsorption behavior of different fragments of EVA molecules can also influence the adsorption behavior of wax molecules on the surface of EVA-attached MCNT nanoparticle, which may influence the effect of C2-100 to improve the rheological behavior of waxy crude oil.

589683 Solar Energy Conversion and Noise Characterization Sensing in Building Integrated Solar Photovoltaic Devices with Ventilation

Abstract:

A study on solar energy conversion and noise characterization sensing in solar photovoltaic devices with ventilation is conducted. A building integrated parallel plate solar photovoltaic (PV) device was built with a pair of PV modules, a ventilated air cavity and an insulating back panel of plywood board filled with polystyrene installed in an outdoor test-room. Photovoltaic amplification was attained with power output from a potentiometer through rotation of its circular knob. A parallel plate PV device was studied for its
electrical parameters as Resistance-Capacitance (RC) electrical analog circuit. The effect of inductive and capacitive heating losses was considered in evaluating electrical characteristics of a PV device exposed to solar radiation. The characterization of noise interference, due to power difference of two intensities in a wave on PV Device is presented. Noise characterization of a wave is obtained depending on type of wave. The difference of two power intensities is due to transmission of light, sound, heat, electricity, fluid, and fire into a parallel plate PV device. Based on analysis of a solar PV device, a definition with measurement expression and unit for an electrical noise were coined. Standard definitions of noise sources, their measurement equations, their units and their origins under limiting reference conditions were also devised. A review for sensing noise of physical agents is conducted for sensing noise of physical agents (sun, light, sound, heat, fluid, electricity and fire). With integrated environmental control, these environmental parameters are to be characterized through various environmental monitoring sensors.

589945 Structural Characterization and Model Representation of Vitrinite-Rich India Coal By Using Advanced Analytical Techniques

Abstract:

The detailed investigation of the chemical structure of Karharbari coal (vitritine rich subbituminous coal from India) by utilizing advanced instrumental analytical techniques such as solid state ^{13}C nuclear magnetic resonance (^{13}C-NMR) spectroscopy, Fourier transform infrared (FT-IR) spectroscopy, X-ray diffraction (XRD), Flash-pyrolysis gas chromatography mass spectroscopy (Py-GC/MS), Laser desorption ionization time-of-flight mass spectroscopy (LD-TOF-MS), X-ray photoelectron spectroscopy (XPS), and high-resolution transmission electron microscopy (HRTEM). The Karharbari coal was an amorphous loosely organized lower rank coal with aromaticity 0.79. This was indicated by XRD results and also confirmed with ^{13}C-NMR results (aromaticity =0.84) and FTIR results (aromaticity = 0.82). The average molecular weight of the aromatic cluster is 507 amu by NMR result which is closer to the result obtained by HRTEM (MW = 530.46 amu). The presence of different heteroatoms like carbon, oxygen, nitrogen and sulphur with their functionalities, were determined by using a XPS technique. Different carbon/oxygen forms present in the Karharbari coal are (C-C) and (C-H) 68.53%, (C-O) 23.46%, (COO^-) 1.96%, (C=O) 6.05% and different nitrogen forms such as pyridine, pyrrolic, quaternary, oxidized and their distribution (mol %) are 19.36%, 45.68%, 31.25%, 3.71 respectively, and different sulphur forms were present i.e. corresponding to thiophenes, sulphones, sulphuric acid and sulphates with the molar content of 16.47%, 41.61%, 21.37%, 20.55% respectively. The construction of the individual molecules will be discussed in detail in the rest of this paper. The individual molecules were geometry-optimized and Materials Studio Amorphous Builder module was used to construct a three dimensional associated structure.

589479 Life Management of High Energy Piping with Highlights from ASME B31.1

Abstract:

High energy piping (HEP) systems – or as defined by the ASME B31.1 Power Piping Code, Covered Piping Systems (CPS) – are critical assets with respect to production facilities. A failure can have catastrophic consequences from a safety perspective and in terms of equipment loss and downtime. In fact, loss of high-pressure steam due to a failure in a main run of steam supply piping could shut an entire production unit down. Because of this and the requirements of the Code, HEP maintenance and inspection programs are established to ensure that the integrity of the systems is maintained in a safe and reliable condition throughout their lifecycle. There is no one-size-fits-all for the life management of these critical assets because there is a great variety in age, design, materials, and owner objectives. Life management of these assets is beneficial in that it provides a structured, systematic approach to creating and implementing a program including evaluating the current program (whether formal or informal), what the objectives of the program are (short and long term), what equipment is in place, and its history.

A HEP program put in place for overall life management is not synonymous with an inspection program, although inspections are an important part of an overall program. Inspections should be used in conjunction with analytical tools, real time monitoring, and laboratory examinations. The overall approach is life management – not inspection management – of the equipment.

This paper describes a HEP life management approach which includes identifying locations of interest, risk-ranking those locations, implementing a plan for determining the health of the highest risk locations, and taking actions based on the system health that has been determined. This paper also presents and discusses some of the requirements of the B31.1 code, which at the highest level states that each operating company shall develop operation and maintenance procedures for piping systems deemed necessary to ensure safe facility operations based on the provisions of the code, relevant industry experience, the operating company's experience (and knowledge of its facility), and the conditions under which the piping systems are operated.

590131 Optimization of C3MR LNG Liquefaction Process Using a Parallelized Simulator

Abstract:

Controlling the cost of energy-intensive processes such as C3MR (propane/mixed refrigerant) LNG liquefaction is a prime opportunity for optimization. Our work combines different C3MR models to develop an optimization that uses multiple independent variables, relaxed constraints, and flowsheet alterations to achieve an optimal operating cost. This work uses a rigorous model of the process in a process simulator rather than a reduced model of the process response. Use of a first-principles process simulator increases the accuracy of the model and therefore allows a more realistic optimization, at the cost of increased computing time. Parallelization (using multiple CPU cores) in the process simulator greatly increases the speed of the optimization calculations. CHEMCAD's optimization tools and multi-core technology make it possible to easily define the optimizations, run the optimizations efficiently, and generate global optima with a high degree of confidence.

590196 Modifying Fischer-Tropsch Reaction Sequence with Addition of Zeolite to Iron-Based Activated-Carbon Supported Catalyst

Abstract:

In addition to previously presented scientific work, modified reaction sequence of Fischer-Tropsch (FT) process was studied by addition of Zeolite ZSM-5 to an iron-based activated-carbon-supported catalyst. Main reactions involved due to FT catalyst are Fischer Tropsch Synthesis (FTS) and Water Gas Shift (WGS) reactions. Main reactions due to ZSM-5 are cracking, oligomerization, isomerization, cyclization, aromatization, conversion of alcohols to hydrocarbons, and coke formation. The waxy and linear structure FT products are converted to gasoline range, lighter hydrocarbons, iso-paraffins, and aromatics using ZSM-5. The catalyst ratio of 1:2, FT catalyst: ZSM-5, is used at 300 PSIG and 300°F. Separate-bed configuration, i.e. FT and ZSM-5 are separated in a reactor, arrangement is used for the reaction sequence analysis. Alcohols with carbon-chain from 3-5 produced by FTS are converted into hydrocarbons over ZSM-5. Waxy hydrocarbons with carbon-chain greater than 23 produced by FTS are cracked and isomerized to small n-paraffins and iso-paraffins using ZSM-5. The hydrocarbon with carbon chain from 7-10 produced by ZSM-5 cracking reactions are converted to aromatic products over ZSM-5. Produced gasoline to kerosene range n-paraffins are further cracked into smaller chain (1-5) iso- and n-paraffins. Some of the iso-paraffins are oligomerized and aromatized over ZSM-5 to form more aromatics. Further, multiple aromatic rings combine to undergo cyclization and dehydrogenation reactions over ZSM-5 to form coke. Water produced as a byproduct of FTS is combined with some of the carbon monoxide, which is the WGS, to form carbon dioxide and hydrogen.

590427 Lateral Retention Force Required for Sliding Saline Droplets over Silianized Silicon Surface

Abstract:

The lateral retention force required to slide a liquid drop on a surface is expected to increase with the time that the drop rests on the surface prior to sliding till it reaches a plateau. The molecular reorientation of the molecules at the periphery of the solid surface subjected to the unresolved vertical component of surface tension force causes this phenomenon. In this system, the surface tension of deionized water is increased by adding NaCl and is expected that the lateral retention force is increased proportionally with concentration. The retention force increased with increase in concentration reaching a maximum only to drop to a minimum with concentrations closer to saturation. That is, the lowest sliding force is required for both lower and higher concentrations of NaCl. This sudden decrease in observed lateral force is not a function of surface tension and the surface on which the drop slides. This phenomenon is associated with adhesion of salt to the surface and is supported by the drop sliding motion and depinning of the liquid from the solid.

590433 Study of Droplet Motion across Polymeric Substrates By Centrifugal Adhesion Balance (CAB)

Abstract:

On a given surface, the contour of the drop changes accordingly based on factors of the force applied, chemical structure of the fluid, drop size, and surface topography. With increasing force that is exerted on the drop, we determine the comprehensive trend for drop motion with respect to the controlled factors mentioned above. Drop motion is categorized into four phases and the point at where the initial displacement of receding line is observed is defined as the onset of motion. In this study, the onset of motion is affected by the factors mentioned above and aids in the determination for another value correlation known as Gs, known as Interfacial Modulus. The resistance to interaction towards the droplet as observed is quantified with respect to change in surface topography. This study therefore characterized drop motion and specifically onset motion based on controlled variables (factors). Work of adhesion is also quantified across different substrates to characterize the detachment of liquid from its substrate. All these parameters such as Gs, onset of motion, contact angles were analyzed with different polymeric substrates, drop sizes, and various fluids used.

590665 Population Balance Modeling and Spatially-Oriented Control for Crystal Size and Shape in Crystallization Processes

Abstract:

Crystallization is an important separation and purification process with relevant applications in the food, pharmaceutical, fine chemical and fertilizing industries. In these processes, modeling and process control for crystallization are focused on crystal size distribution (CSD) and shape of the crystals in order to make it possible to achieve the desired quality and purity of the product of interest. The control of crystal size and shape is of considerable importance for industrial crystallization processes, as crystal morphology has great influence on downstream steps and also from the point of view of product specification criteria. The consideration of crystal shape in modeling of crystallization processes is also important due to its impact on product effectiveness, such as the bioavailability and tablet stability in pharmaceuticals. Control of the crystallization process is can be facilitated with online measurements, which provide information to be used in the modeling of the process. However, if online measurements are not available, the strongly nonlinear dynamics of these processes makes difficult to establish relations between some inputs (e.g. temperature) and the final desirable characteristics of the crystal product. In this context, image analysis has been shown to be a suitable technique to monitoring the process [1].

In our present work, a population balance model-based scheme for controlling the size and shape of crystals is presented. The modeling was performed using a two-dimensional population balance (2D-PBM) in terms of the two characteristic dimensions of the crystals, describing the evolution of crystal size and shape distributions along the batch time. To identify the nucleation, growth and dissolution kinetics parameters using the 2D-PBM approach along the length and width directions, QICPIC-LIXELL's high speed dynamic image analysis equipment was used for the online measurement of the CSD, shape parameters and aspect ratio distribution of crystals during the experiments. In this way, the developed model allowed for the prediction of the crystal size and shape distributions during growth-dissolution cycles.

The developed framework was employed to obtain optimal temperature policies for controlling the crystal size and shape distributions that meet target criteria for size, morphology and productivity using the dynamic image analysis along the runs. Based on the concept of using measurements along batch crystallization processes as a trajectory in a phase space [2], a trajectory-endpoint control problem is presented with an oriented control scheme that is applied to the shape of the crystals. The scheme was used to control the batch cooling crystallization of potassium dihydrogen phosphate. The reported experimental data demonstrate the application of the control policies for temperature to produce crystals of the desired average shape and size in different batch times. Thus, using the first principle modeling coupled to the developed trajectory tracking control, the proposed framework has the potential to be applied to industrial crystallization processes for shape control and to be extended to polymorph control investigations.

References:

[1] Eisenschmidt, H.; Bajcinca, N.; Sundmacher, K. (2016) Optimal Control of Crystal Shapes in Batch Crystallization Experiments by Growth-Dissolution Cycles. *Crystal Growth & Design*, v. 16, p. 3297-2803.

[2] Griffin, D. J., Grover, M. A., Kawajiri, Y., Rousseau, R. W. (2015) Mass-count plots for crystal size control. *Chemical Engineering Science*, v. 137, p. 338-351.

590450 Auditing the Process Hazard Analysis - Key Characteristics to Scrutinize in a Good-Quality HAZOP Study and Supporting Effective HAZOP Study Updates

Abstract:

Although Process Safety Management (PSM), requiring good-quality Process Hazard Analyses, was promulgated in the United States in 1992, and HAZOP Study techniques date back to the mid-1970's, the HAZOP Study technique is still not uniformly applied in a high quality fashion. The objective of this paper is to provide experience-based tips on Process Hazard Analysis auditing, and some key characteristics that the auditor can scrutinize to assure quality implementation. Analogously, a similar set of characteristics that can facilitate future HAZOP Study updates can also contribute to a quality HAZOP Study that supports the spirit of Safety Management System objectives for the identification of hazards potentially impacting plant personnel, the community, and the environment. This paper will also review tips for structuring the HAZOP Study to facilitate later updates and revalidation.

590342 Development of High Performance Organic-Based Cathodes

Abstract:

For lithium-based batteries, structure/property relations in terms of cell voltage are desired for new organic-based cathode active materials. In this initial pursuit of structure/property relations, ab-initio calculations were performed on a few organic-based molecules to show the effect on the theoretical cell voltage by inclusion of nitrogen and/or oxygen. Preliminary test results are presented on a newly developed molecule and methods to potentially improve on the cell performance are given.

590705 Electrochemical Kinetics of Degradation of Graphite Anodes in Lithium Ion Batteries

Abstract:

Growth of the solid electrolyte interphase (SEI) is a major driver of capacity fade in LIBs. Despite its importance, the fundamental mechanisms remain unclear, primarily because of the complicated reaction pathways involved[1–3]. SEI growth can be both electrochemical and chemical in nature[4], and thus, it is a strong function of the potential and degree of lithiation of the electrode. We model the early-stage and long-term growth of SEI by accurately capturing the potential dependence of its formation kinetics as well as long term rate limiting steps, and validating it against the world's largest open source battery cycling data, generated in-house[5]. This is done using the Multiphase Porous Electrode Theory (MPET) framework[6] on graphite (phase separating) and carbon black (non phase separating) particles.

Lithium plating is another key degradation phenomenon that has been elusive, and it becomes important while trying to fast-charge batteries, i.e., 0% - 80% state-of-charge in 30 mins. We show that lithium plating is a key function of electrode morphology, phase-separation dynamics and potential. Phase-separation in graphite is modeled in the electrode using the Cahn-Hilliard Reaction framework described by Bazant[7]. We understand the electrochemistry of the onset of lithium plating with in-situ measurements connected to real time cell potential in a phase-separating electrode for the first time[8].

Results indicate that the peak SEI-forming currents are higher for higher driving currents. Also, we find that SEI only grows during electrode lithiation, i.e. the battery only degrades while being charged. We also find that onset of lithium plating is correctly captured only when phase separation in active material is accounted for. Further, the onset of plating is delayed on electrodes with a thick SEI layer – understanding SEI/plating coupling is integral to predicting fast charging manufacturing protocols for LIBs. This work holds promise for the predictive design of procedures[9] for manufacture and formation of LIBs.

[1] Cohen, Y. S.; Cohen, Y.; Aurbach, D. Micromorphological Studies of Lithium Electrodes in Alkyl Carbonate Solutions Using in Situ Atomic Force Microscopy. J. Phys. Chem. B 2000, 104, 12282–12291. https://doi.org/10.1021/jp002526b.

[2] Horstmann, B.; Single, F.; Latz, A. Review on Multi-Scale Models of Solid-Electrolyte Interphase Formation, Current Opinion in Electrochemistry 13, 62-69 2019.

[3] Nie, M.; Abraham, D. P.; Seo, D. M.; Chen, Y.; Bose, A.; Lucht, B. L. Role of Solution Structure in Solid Electrolyte Interphase Formation on Graphite with LiPF6 in Propylene Carbonate. J. Phys. Chem. C 2013, 117 (48), 25381–25389, https://doi.org/10.1021/jp409765w.

[4] Das, S.; Attia, P. M.; Chueh, W. C.; Bazant, M. Z. Electrochemical Kinetics of SEI Growth on Carbon Black: Part II. Modeling. J. Electrochem. Soc. 2019, 166 (4), E107–E118. https://doi.org/10.1149/2.0241904jes.

[5] Attia, P. M.; Das, S.; Harris, S. J.; Bazant, M. Z.; Chueh, W. C. Electrochemical Kinetics of SEI Growth on Carbon Black: Part I. Experiments. J. Electrochem. Soc. 2019, 166 (4), E97–E106. https://doi.org/10.1149/2.0231904jes.

[6] Smith, R. B.; Bazant, M. Z. Multiphase Porous Electrode Theory, J. Electrochem. Soc. 2017, 164 (11). https://doi.org/10.1149/2.0171711jes.

[7] Bazant, M. Z., Theory of chemical kinetics and charge transfer based on nonequilibrium thermodynamics, *Accounts of Chemical Research*, *46*(5), 1144–1160. https://doi.org/10.1021/ar300145c

[8] T. Gao, Y. Han, S. Das, T. Zhou, D. Fraggedakis, N. Nadkarni, C. N. Yeh, W. Chueh, J. Li, M.Z. Bazant, Interplay of lithium intercalation and plating on graphite using in-situ optical measurements, submitted.

[9] Huang, W.; Attia, P. M.; Wang, H.; Renfrew, S. E.; Jin, N.; Das, S.; Zhang, Z.; Boyle, D. T.; Li, Y.; Bazant, M. Z.; McCloskey, B. D.; Chueh, W. C. and Cui, Y.; Nano Letters 2019 19 (8), 5140-5148. DOI: 10.1021/acs.nanolett.9b01515

589787 Model Development of Solid Oxide Fuel Cell Thermal Performance Using Artificial Neural Network

Abstract:

NASA's Johnson Space Center has recently begun efforts to eventually integrate air-independent Solid Oxide Fuel Cell systems, with landers that can be propelled by LOX-CH4, for long duration missions. Using landers that utilize such propellants, provides the opportunity to use such systems as a power option, especially since they will be able to process methane into a reactant through fuel reformation. To ensure fuel reformation in the systems, Steam Methane Reformation (SMR) will be employed. Various lead-up activities, such as hardware testing and computational modelling, have been initiated to assist with this developmental effort including the development of dynamic empirical models using artificial neural network. Such a modeling approach has shown to be accurate ($R^2>0.9$) to predict the thermal and electrical performance for SOFC systems. Hence, such models could be used for system optimization and control design.

Impact of Spacer Nature on Rheological Properties of Cationic Gemini Surfactant

Abstract ID: 589650

Shams Kalam[1], Muhammad Shahzad Kamal[2], Shirish Patil[1] and S. M. Shakil Hussain[2]

(1) KFUPM, Dhahran, Saudi Arabia

(2) Center for Integrative Petroleum Research, King Fahd University of Petroleum & Minerals, Dhahran, Saudi Arabia

Surfactants play a vital role in Chemical Enhanced Oil Recovery (cEOR) techniques. Gemini surfactants are a unique style of surfactants having more water-soluble hydrophilic head and oil-soluble hydrophobic tail units. Both groups are connected by a spacer at or near to the head groups. Polymers are added into the surfactants to increase the sweep efficiency of the flooding processes. This paper presents the rheological properties of Surfactant Polymer (SP) solutions in which two recently synthesized cationic gemini surfactants with different spacer nature are used.

The rheological properties of novel SP solutions were determined using Discovery hybrid rheometer (DHR-3) from TA Instrument. The geometric configuration set for this work was concentric cylinder. All experiments were carried out at 80 °C, while the shear rate for steady shear viscosity measurements ranges from 0.001 to 1000 s^{-1}. 0.25 weight % of a polymer FO 4290 SSH was used with different concentrations of newly synthesized gemini surfactants for Rheological measurements. Two hours magnetic stirring and twenty hours retention time were provided to make uniform surfactant polymer solutions.

Both surfactants exhibited good thermal stability in seawater with no precipitation. Shear viscosity and storage modulus were measured as a function of shear rate and angular frequency respectively at different surfactant concentrations ranging from 0 to 0.5 mM, where 0 mM shows pure polymer. We discovered that counter ions also influence the rheological properties significantly. Research outcomes revealed that the novel SP solution with a benzene ring and chlorine counterion produces better performance in comparison with benzene ring and bromine counterion. Moreover, the effect enhances if we increase one more benzene ring in the spacer.

This study reports two recently synthesized SP solutions for cEOR processes in carbonate reservoirs. It presents the effect of counterion on rheological properties and improves the understanding of tuning the surfactant structure in making SP solutions to increase the flooding sweep efficiency.

590257 Real Time Energy Management Systems Implementation for Continually Optimized Performance

Abstract:

It is well known that energy management is key for the optimal operation of refining and petrochemical sites. Industrial energy systems are constituted by complex utilities networks that are interconnected and have complex interactions that in order to achieve it best performance need to comply with increasing environmental rules, push against operating constraints and adapt to altering power prices and rapidly changing equipment availability.

After the initial installation, management and continuous improvement of these systems present challenges not only from a technological but also from an organizational point of view.

The implementation of an advanced digital Energy Management System (EMS) which includes capabilities for multi-period optimized scheduling, real time optimization and monitoring, deployed either as cloud-based or on-premise solutions that could be acting on open or closed loop is known to provides large benefits.

Critical is the sustainability of the EMS in the ever-evolving energy supply and demand of refining and petrochemical sites. Commercial, asset and organizational factors all change and the EMS needs to adapt to the developing environment of each enterprise meaning the type of implementation, including appropriate management structure, culture, best practices and Key Performance Indicators (KPIs) needs to be continually improved.

The EMS implementation at the large and complex Eastman site in Longview, TX, will be presented as an example. It will provide insight on the initial successes achieved in real time energy management for both, cost optimization and energy related KPIs performance tracking. The case study evolves to discuss how technical asset improvements, once replicated in the EMS, provided alternative operating modes, large additional benefits and sustained optimal

Andrew Lammers, NN2

Eastman Longview

Pablo Montagna*, Cristian Rojas

KBC Advanced Technologies, Inc

589818 Analysis of Fuel Cell System through Project Based Learning for Application of Classical Thermal Fluid Concepts in Mechanical Engineering Laboratory Course

Abstract:

A hands-on project involving analysis of air noncircular channels in Fuel Cell to measure the pressure was introduced in a thermal fluid laboratory course. Although the students studied fluid mechanics as a theoretical course, this project gave them the opportunity to build an experimental setup from scratch to help showcase their hands-on skills and to apply their understanding of relevant concepts of fluid sciences. The results of the project included a successful test of a conceptual procedure, designed to measure and calculate pressure drop and flow rates across the anode and cathode sides of a fuel cell. The students enjoyed completing and learning from the project. They also provided very useful feedback on the project by incorporating a functional experimental system for the future laboratory project. This project based learning opportunity allowed students to build their confidence in what they are learning as well as provide a glimpse of what the real world may hold.

589438 Measured Versus Calculated Sour Gas Properties

Abstract:

Measured versus calculated sour gas properties
Abstract:
Recent Environmental agencies have restricted standards for sulfur emission from gas chemical processing. Sour gas can extremely impact the delivery and processing by corroding the equipment. Gas properties such as compressibility, density and viscosity are a major factor controlling prediction of future production pipelines, and production tubing and pipelines in gas industry.
Sour gas effect on all gas industry aspects usually gas properties are measured in laboratories but occasionally experimental values are not available due to complexity and time consuming hence a reliable method would be proposed.
We are assessing the accuracy of the industry standard methods in calculating sour gas properties, compressibility factor using the following correlations:
Drunchuck - Abukkasem, Hall – Yarbarough, Drunchuck- Purvis- Robinson, Londono et-al, Alanazi et-al and Azizi et-al.
Viscosity using empirical correlations such as Carr-Kobayashi Burrows, Lea-Gonzalez –Eakin and Dean- Stiel.
Density using equations of state Redlich Kwang and Soave -Redlick - Kwang.
This study shows that the best correlation for calculating compressibility factor and density of high concentration sour gas is Londono-et al, all other considered methods are not accurate. All other methods investigated include high error especially for high H2S concentration.
However the error in estimation of Z factor for most of the considered methods has great impact on estimated sour gas density. Therefore we recommend that sour gas properties be measured experimentally at high pressure and high temperature conditions.

587967 The Study of a S-CO2 Oxy-Fuel Combustion Power Plant for Low Grade Coals

Abstract:

Recently, it is very important to have reliable solutions that can maximize power generation efficiency of current power generation processes and decrease emissions of CO_2 and other pollutants because there are increasing concerns about the global warming and the depletion of fossil fuels. Among the available carbon capture technologies and power generation technologies, a S-CO_2 (supercritical carbon dioxide) cycle based oxy-fuel combustion process appears to be a promising mid-term solution for the production of electricity from fossil fuels. Since supercritical carbon dioxide (S-CO_2) involves a rapid density change at small temperature and pressure ranges, the S-CO_2 cycle can extract large amounts of energy from relatively small equipment and it is possible to minimize or simplify a power generation block and to obtain higher thermal efficiency than existing steam cycles used in power generation plants. Therefore, an indirect S-CO_2 oxy-fuel combustion power plants for low grade coals was designed and analyzed by using ASPEN-PLUS. Also, the operating characteristics of the indirect S-CO_2 Oxy-fuel combustion power plant were compared with those of conventional air combustion power plants.

590099 Characterization and Performance Assessment of Oil Entrapment By Commercially Available Surfactants

Abstract:

Oil spills occur frequently and are a threat to the health and safety of human society and the well-being of the environment and ecosystem. Remediating oil spills is an ongoing challenge. Surfactant based dispersing agents, such as the Corexit® product line, are widely used for remediation of oil spills mainly through direct application on top of the oil slick. This work encompasses the investigation of variables that influence the emulsification process and therefore the encapsulation of oil within the surfactant micelles. Of central importance are the conditions under which oil entrapment in emulsions even occurs. A central finding is that there appears to be a threshold shear rate for emulsifying oil with surfactants, below which no emulsification occurs. Different schemes for applying shear stress in this work included simply stirring, using a high energy colloid mill, and flowing the mixture through a high shear capillary. Also, the sequence in which the oil/surfactant/water (O/S/W) components are mixed impinges strongly on whether and how much oil entrapment occurs. To try to correlate shear stress and velocity profiles throughout the sample volume with entrapment efficiency and kinetic stability of these O/S/W systems, Computational fluid dynamics (CFD) simulations were conducted. Another important factor is the ionic strength of the aqueous medium, which greatly impacts kinetics of these emulsions, but with a smaller impact on the threshold at which any emulsion is obtained at all. Surfactants tested include Corexit® EC9500A, sodium dodecyl sulfate (SDS), Triton X-100, and cetyltrimethyl ammonium bromide (CTAB). The quantification of oil entrapment was performed through either thermal gravimetric analysis (TGA) or differential scanning calorimetry (DSC).

Designing a Native Model for Assessment of the Effectiveness of Advertising

Sorush Niknamian
PhD, military Medicine Department, Liberty University, USA
Email: so.niknamian@gmail.com

Abstract

Reviewing and evaluating the effectiveness of advertising can be used as a criterion to convert weaknesses into strengths and budgeting. Also, Identification of the factors influencing the effectiveness of advertising can lead to more professional design and implementation. In the banking industry, with the increasing number of banks and increased right of customers to choose, the role of advertising has become increasingly important to succeed in this field of competition; so it should be said that advertising can make the bank close to its goals when it is sufficiently effective. The current research aimed to identify and rank the effective factors on the design of a native model for evaluation of the effectiveness of advertising of the Eghtesad Novin bank in Iran. The statistical population of this research is all managers and experts of the Eghtesad Novin bank. According to the research literature, the criteria and sub-criteria affecting the effectiveness of advertising have been extracted. By using the fuzzy Delphi technique and based on the experts' opinion, the final indices were extracted. Then, using the fuzzy AHP method, the rank of criteria has been addressed. The derived criteria are demographic characteristics, individual emotions, and attitudes of the customer, apparent features of the message, message content, market characteristics, cultural, economic features, design and nature of the service, advertising methods. The results of the ranking showed that the methods of advertisement, the design, and nature of the service, the apparent characteristics of the message, the emotions and attitudes of the customer, the demographic characteristics, the content of the message, economic, market characteristics, were placed in the first to ninth ranks, respectively.

Keywords: banking services, banking services marketing, advertising, advertising effectiveness,

Introduction:

Advertising is one of the promotional mix elements of marketing mixes, which play an important role in attracting customers. Advertising is a bridge between the organization and customers and can introduce the organization and its products to customers, inform consumers about the goods and services of the organization, convince them about the benefits of the product, and ensure them to have the best choice even after the purchase when the customer is uncertain (Buck, 2016: 172).

Given the increasing competition among the brands in the marketing sector, advertising is a necessary tool to introduce the characteristics of goods and services to customers. Leong et al. (1999) pointed out the fact that consumers overcome advertising information, and many of them do not remember the advertisements that they are exposed to, and it is necessary to consider the advertising and its attractive methods to attract customers' attention. The ability to attract customers' attention is one of the main indicators of advertising effectiveness that should be taken into consideration. The attention of customers is a key component of the overall effectiveness of advertising (Pileliene et al., 2017: 2). Any professional marketing without advertising will not be effective. Despite relatively high advertising costs, advertising is mentioned as an investment rather than the cost. Since many advertisements cause that the advertising companies do not benefit their advertising because of the wrong way of advertising, the competitive companies benefit from the advertising of that company (Hartnet, 2016: 23)

Advertising should be in line with the goals and plans, as well as the strategies of the organization. Therefore, we should see to what extent the advertising has met such goals. Allocating funds without evaluating the effectiveness of past advertising is the waste of resources of the organization. By evaluating the effectiveness of advertising, changes can be made to the form, the content of messages, type of media, budget, communication channels, and even the time and conditions for conducting advertising to create more useful and effective advertising than the past (Yazgirudo-Yasteh, 2015: 359). In the current competitive world, the importance of advertising is increasing. As a result, businesses have devoted a great deal of money to customers' attraction, but it is important to see whether the company advertising has been designed and implemented in such a way to meet the goals of the company. This issue has been often neglected in our country. (Kim, 2014: 2107).

The amazing effect of advertising on the lives of individuals requires that this scientific, social and communicative branch be carefully studied and advertising is made with a scientific, artistic, specialized and expert view. Professional advertisement can be a sharp sword for sales promotion, both inside and outside the country, and otherwise, it is a waste of time, money and energy. A professional advertising lever can, with the help of new technology, penetrate humans, and cause more desire for consumption (Arribeh, 2016: 463). Today, banks have turned their attention to the widespread use of mass media advertising, given the highly competitive structure governing on them. Also, given the relationship among a wide range of businesses of organizations and households and banks through various deposit accounts, loans, and other financial services, banks can benefit from advertising in a proper way to meet their goals, namely the introduction of their services, the communication with customers, their effectiveness on them in choosing a bank to do their banking business and ultimately attracting them. Unfortunately, in most banks, despite the massive advertising and the allocation of large amounts of energy and budget, the effectiveness of such advertisements is less evaluated. Since advertising is one-way, it is necessary to receive feedback from the customers (Rabiei, 2011: 19).

Regarding marketing, the effectiveness of advertising depends on the wide range of coverage of various consumers and the cost of advertising campaigns. Economists have defined the effectiveness of advertising as a change in the profit derived from promotional activities. Therefore, the effectiveness of bank advertising is the potential quality of information reported to consumers, and in promotional campaigns, it aims at creating a bank image and informing banking products and services, as well as obtaining a specific economic result before and after the transmission of the message (Denis et al., 2012: 85).

Eghtesad Novin Bank (EN bank), which is one of the oldest banks in Iran, due to the addition of new banks, should use new tools to protect and even increase its market share for the assessment of the effectiveness of its advertising. For this purpose and given that the bank uses various advertisements, it is necessary to design a model to evaluate the effectiveness of the bank advertising; therefore, the present study seeks to answer the following questions:

- What indicators are used to assess the bank native advertising model?

- How are the indicators for bank advertising evaluation weighed and ranked?

In this research, we have tried to identify the criteria using the Fuzzy Delphi process; then we will use the Fuzzy AHP method to rank the identified criteria. In the next section of the research, the literature has been studied. Then the research design has been expressed, and in the last section the conclusion has been presented.

1-Research literature:

1-1-Marketing of banking services

The growing trend in service activities has led to the emergence of new markets and large service companies. The complexity of service industries, such as information systems, management consultancy, and logistics, has increased, as service providers have entered the market with a variety of business backgrounds and capabilities to provide a broader range of services. Facing this complexity requires knowledge and skills to understand the services. The definition of the service includes all activities that have been done (for example, writing document specifications) (Selviaridis et al., 2013: 1398).

In the industrial literature, there are various ways to define the service and product requirements. Van Weele (2010) has distinguished technical and applied definitions from each other. The first requirement expresses the exact technical specifications of the goods with the required services and inputs while it focuses on performance demands and demand-oriented performance orientations. In this regard, Axelsson and Wynstra (2002) have proposed four methods for definition of services: input definitions are based on the sources and capabilities of the service provider while the process-oriented approach focuses on how services are presented regarding activities and processes. Output-oriented definitions focus on service capabilities and performance levels, as well as the interaction between the suppliers and recipients. Ultimately, they focus on stress-oriented results in the economic value of services (Selviaridis et al., 2013: 1399).

The financial sector forms an essential part of the service sector. In this regard, banks have used marketing to improve the strategic performance of banks. In particular, since the 1980s, this trend has been confirmed and criticized. Marketing researchers and experts are interested in researching the various aspects of marketing in banking. The quality of services is essential for customers who use banking services, and the provision of accurate and high-quality services leads to the survival of banks in the competitive market. Improving the quality of services will increase the access and response to the demands of the bank's customers, which will make the bank more distinct than other competitors. Therefore, measuring the quality of services in the banking industry is essential. Providing quality services is a hard and difficult task, in this regard, the features of services in general and banking services, in particular, should be explained. Banking services require a high degree of appropriate communication between service providers and customers. If services are not provided properly, they will be at high risk.

About the five features that distinguish them from material services (intangibility, non-allocation of materials for production, diversification, non-corruption, and lack of ownership), financial services have two other specific features, that have been firstly developed by McKechnie (1997). These characteristics include trusteeship and two-way information. These special features mean that the evaluation of services is more difficult than material products, and they not only create a challenge in the development of marketing activities, but also they are the source of the problem in providing an optimal level of quality services. Banking services require a high level of communication between service providers and customers, as well as a high degree of customer participation in the process of providing services. Angurr et al. (1999) emphasized that due to the high standardization and the development of technologies in the banking sector, human factor has been considered as the most important one in making distinctions among competitors. The expertise and motivation of the employees, as well as the level of customer involvement in the provision of services, significantly affect the quality of services. Due to the wide range of banking activities, attention should be paid to cultural and social changes that affect the attitudes of customers (Durdana et al., 2015: 100).

Marketing of banking services has three components as follows: if any of them were properly executed, banking services would change the country.

1- Market study: Identification of the banking services market system, including the status of current banking services and their weaknesses, strengths, expectations, demands, ideas and opinions of customers, the conditions of environmental instruments affecting the market system, in particular government policies and governing laws, and even global changes in the banking system and successful models in this system. Indeed, market study is the foundation of the evolution of the banking system, because market study (which is the result of constant communication with the market and conducting marketing studies and research in various forms) leads to comprehensive, up-to-date, relevant and necessary information for decision-makers and bank managers (Munsung Rheea & Satish Mehrab, 2006: 508).

2. Market creation: It involves the use of a variety of simple and complex marketing techniques and competitive strategies, as well as the growth, development and market creation in the banking system; it needs to understand the internal customers (employees) and external

customers, to communicate with them and meet their needs and demands. Market creation in the competitive environment requires appropriate strategies (Moch, 2013: 2908).

3. Market protection: One of the most important tasks of marketing in the banking system of the country is to maintain market share, a stable relationship with customers, and to convert ordinary customers to loyal and supportive ones. The more intense the competition, the more valuable the loyal customers will be, as marketing research in the world has shown that the cost of finding a new customer in firms is 3 to 33 times higher than the cost of maintaining current customers regarding the type of products and services. Therefore, market protection prevents the costly advertising, and attraction of new customers and provides a kind of confidence and serenity for market survival, especially in competitive conditions (Fredriksson & Moro, 2014: 69).

1-2-Effectiveness of bank advertising

One of the indicators of success in any organization is the extent to which the organization achieves its goals. Here, the success of any organization is determined by its own goals. This approach has been called effectiveness in the management literature. The effectiveness is the degree to which the organization achieves its goals. Goals must first be defined to measure effectiveness in any organization. Then, plans are set up to fulfill these goals, and thus the plans will be implemented. In the end, the results are compared with predetermined goals (Manjegani et al., 2015: 124).

About 100 years ago, William Heketh statement, "Half of the advertisements were wasted, we just do not know which half it is" attracted everyone to the accountability of advertising. What matters to advertisers is that how much the amount of money spent on advertising has returned? Has the cost spent on making and broadcasting the advertising been effective? Has the effect of advertising been strong or weak? Knowing the answer to these questions is important to the advertiser because today companies pay much money to make and broadcast advertising. According to Jack Trout, the money we use to advertise is very much. Every second of the World War II was $ 9,000, and Vietnam cost $ 22,000 per minute. But in the American Football Final, every minute of advertising cost $ 2 million, so the advertiser has the right to know about the effectiveness of his advertising (Aksoy, Atilla, 2005, Yeni Reklamcilik, Istanbul Bilgi Universitesi, 95). It is very difficult to measure the impact of long-term advertising on many occasions. Also, academically in the world of advertising, it is impossible to create a comprehensive theory that can be used to determine the effectiveness of advertisements on people at different levels under different circumstances. The effectiveness of media is its capacity to create desired results, or special conditions in a particular position and content, that is, attaining the desired communication goals (Wery, 2012)

Evaluating the effectiveness of advertising indicates how much advertising has succeeded in achieving the goals. The evaluation of the effectiveness of advertising acts as a feedback, informs us about the weaknesses and strengths of our advertising by which we find our shortcomings in the advertising. The marketer may choose an appropriate stage regarding product life or brand position, and use it to define his advertising goals and ultimately evaluate its effectiveness (Eckler & Bolis, 2011: 2).

The effectiveness of advertising is a function of its content (message), the execution (how it is transmitted), and repetition (the number of times each consumer sees the message) (Cutler, 1997, quoted by Moorthy & Hawkins, 2005). Some of the effectiveness criteria are the call and recognition of the brand name (Park, 2006), awareness, willingness and intention to purchase (Decrop, 2007), attitude towards the ad, attitude toward the brand name and intent to purchase (Muda et al., 2014)

The control of the communication element has a direct effect on customer perception of different dimensions of the brand name, such as service quality and loyalty. It means that the elements of communication, advertising and promotion must be carefully implemented and controlled (Suvittawat, 2015). Informative advertising is one of the factors that shape the behavior of the bank customers. Studies conducted both on attracting, retaining and loyalty of customers, show the need to examine the impact of advertising on attracting and loyalty of customers (Ioannou, Boukas and Skoufari, 2014: 25).

An ideal model of world-class advertising can: 1. transfer a consistent meaning to the consumer all over the world. 2. Transfer positive global capacity for the brand. That is, the results of advertising among different peoples should produce a positive effect and transfer high-quality signals (Strebinger et al., 2018: 410).

Advertisers want to prepare consumers to like goods or services, but sometimes it harms them through unwanted and intrusive advertising. It leads to the creation of a series of negative emotions in advertising. The purpose of ads is to create positive feelings in consumers about advertised goods and services (Bell & Buchner, 2018: 1).

On the other hand, research shows that commercial advertising can be positively evaluated and considered, and this positive evaluation leads to a positive evaluation of the brand with the product. However, advertising can have negative effects that lead to the formation of negative attitudes towards that product and service. Therefore, by increasing the volume of messages, it can have positive, negative or neutral effects. Therefore, marketers should evaluate advertising messages and their behavioral outcomes. With this regard, the effectiveness of advertising, based on a hierarchical approach, should be fully investigated (Tafani et al., 2018: 125).

There are various factors which influence effectiveness of advertisement like media, time, market life stage, advertisement life stage (new or old advertisement) and also advertisement creative signs (like content and administrative elements) (Talis et.al, 2005). Also using popular cartoon personages (Ranjbarian, 2009), size (Rousat, Venous & Ebrahimi, 2004, Baltas, 2003), color (Park, 2008, Moriarty, Mitchell, Wells, 2009, Cheng & Kao, 2011), complexity or simplicity of the advertisement (Morrison & Dainoff, 1972, Di Res, 2008), using irony factor (Mukherjee, Hober, 2004), using famous artists (Rousta, Venous, Ebrahimi , 2004) and consumer features (Matrin, Phaymi, Agee, 2002, Park, 2008) are among these variables. Advertisement repetition is also a factor which helps in reminding its content to the audiences (Rousta, Venous, Ebrahimi, 2004). One of considered factors in the debate of evaluating effectiveness of advertisement is the attitude towards advertisement. Attitude towards the advertisement includes public interest or lack of interest in an advertising stimulus in encountering a special topic (Moaen & Minor, 2009). Mental imagery (B one & Elen, 1992), content transparence and emotions called through the advertisement are among variables influencing the attitude towards the advertisement. Some of these factors may lead to strong impacts while some of them may create weak impacts or don't include any impact. Awareness of the impact of each of these factors makes a considerable help in making management decisions (Talis et.al, 2005).

Fotini et.al (2009) provided a model of effective advertisements for advertising websites of big companies. The results of the study showed that measurements, activities, priorities and intentions of people have a considerable impact on evaluation and qualitative factors of relation-oriented marketing have a great impact on advertisements effectiveness indicators. The extracted criteria include:

- Advertisement policies: product information, company information, product general display, consistency with traditional advertisement

- Medium features: message features, information structure, responding time, facilitating research, beauty, interactions, communications in real-time, amusement, information control

- Relation- oriented marketing policy: communications, feedback, supporting customers

Today, organizations for competition in the business world and selling their goods and services require purposeful and efficient marketing activities. One of topics which has a very important role in the process of marketing of an organization and assign a high volume of budget to it is advertisement. Advertisement causes some changes in the knowledge, attitude and behavior level of audiences, so attention to evaluating advertisement effectiveness is a necessity for the organizations that consider advertisement as requirement of continuous life, growth and development. By assessing advertisement effectiveness and investigating consistency with the organization objectives and its expected impacts, some changes may be created in its dimensions, content and methods so that a more effective and useful advertising is performed. Regarding the research literature and internal and external background which has been stated, here a general categorization of the model proposed criteria has been provided in the following table:

Table 1- summation and extracted variables

scholar	sub standard	criterion
) Dargi & Mohammadian (2010), Ashouri (2016), Mansouri Moayed (2005), Aghapour (2016), Mirzadh & Mehrmanesh (2016), Keramforn (2011), Mac Ki- Nesbiat et.al (2011), Rosemary et.al (2012), Wang (2015), Maryion et.al(2017)	awareness of sexuality differences	individual factors
Dargi & Mohammadian (2010), Ashouri (2016), Samsam Shariat Keramforn (2011)(2007), Mirzadeh & Mehrmanesh (2016), Mac Ki- Nesbiat et.al (2011), Rosemary et.al (2012), Wang (2015), Maryion et.al(2017)	attention to age requirements	
Dargi & Mohammadian (2010), Dehdashti Shahrokh (2009)	various consumers (permanent, ordinary and case consumers)	

Khashizadeh (2010), Danaei & Dasar (2016), Pillay et.al (2012), Marion et.al (2017)	advertisement richness	
Danaei & Dasar (2016), Dargi & Mohammadian (2010), Pillay et.al (2012), Olson et.al (2012)	advertisement accuracy	
Ashouri (2016), Rabiei (2011), Taeini Harari (2014)	positive mentality of the organization	
Dargi & Mohammadian (2010), Shahmohammadi (2011), Olson et.al (2012)	honesty feeling in advertisement	
Fakhari et.al (2012)	threatening feeling	
Ashourizadeh (2016), Hamidizadeh (2013)	attitude towards brand	
madhoushi (2013), Olson et.al (2012)	being convinced	
Ahmadi (2013), Mac Ki – Nesbiat et.al (2011), Olson et.al (2012)	convicting audiences	
Ashouri (2016), Hosivar (2014), Olson et.al (2012)	mental imagery	
Siadat & Sakhi Jourashk (2014)	positive feeling	
Navaei et.al (2015), Rozmari et.al (2012)	audience being active	
Mosavi Esmaeilzadeh (2015), Olson et.al (2012)	knowledge level	
Dargi & Mohammadian (2010), Ashouri (2016), Shafiei Nikabadi & Zarei (2017)	attitude towards advertisement	
Dargi & Mohammadian (2010), Hosseinzadeh et.al (2016), Sanei et.al (2016)	familiarity with methods and forms of advertisement	message
Ashouri (2016), Olson et.al (2012), Alan (2012), Patric Ro (2014)	attention to reduction of expectation time of the user for observing advertisement	
Danaei & Dasar (2016), Forentebury & Mac Gldric (2011)	beautiful and appropriate	

	combination of colors	
Aghapuor (2016), Sanei et.al (2016)	site and location of advertisement placement	
Hassanzadeh & Danaei (2016), Sanei et.al (2016)	access to sufficient and required information	
Dargi & Mohammadian (2010), Darzban (2016)	starting the advertisement with a challenge	
Navaei et.al (2015), Salimi Fard & Pourdehghan (2016), Bargar et.al (2016)	starting the advertisement with a statement and explanation	
Dargi & Mohammadian (2010), Mirzadeh & Mehrmanesh	using helping words and terms for decision making	
Dargi & Mohammadian (2010), Daneshvar et.al (2016), Alan (2012)	using stimuli with diversions more than usual and expected extent	
Sanei et.al (2016)	using images of women and children	
Sohrabi Ranani (2016), Fortenbury & Mac Goldric (2011)	creativity in the message	
Shafei Nikabadi & Zarei (2017), Denis et.al (2012)	the text of advertisement message being a predicate sentence	

Khashizadeh (2010), Dargi & Mohammadian (2010), Sohrabi Ranani (2016), Patric Ro (2014), Abdolsalam et.al (2016)	continuation and repetition of message	
Daneshfar & Nabizadeh (2015), Fortenbury & Mac Goldric (2011)	using trustable and famous people	
Salimifard & Pourdehghan (2016)	common dialects and terms	
Daneshfar & Nabizadeh(2015), Forentnburi & Mac Goldric (2011)	loveliness of the popular person for the observer	
Daneshfar & Nabizadeh (2015), Forentnburi & Mac Goldric)(2011	familiarity of famous person with the observer	
Mansouri Moayed (2005), Aghapour (2016)	emotional music	
Khonsazadeh (2010), Hosseinzadeh et.al (2016), Aghapour (2016)	using various methods	
Siadat & Sakhi Jourashk (2014), Sohrabi Ranani (2016)	high coverage and being universal	
Navaei et.al (2016), Mosavi Esmaeilzadeh (2015), Hosseinzadeh et.al (2016), Mirzadeh & Mehrmanesh (2016), Rosemary et.al (2012), Mikit et.al (2012), Spenkar (2013)	media advertisement	
Navaei et.al (2015), Mosavi Esmaeilzadeh (2015), Hosseinzadeh et.al (2016), Ashouri (2016), Hassanzadeh & Danaei (2016), Mikit et.al (2012)	multimedia advertisement	
Navaei et.al (2015), Mirzadeh & Mehrmanesh (2016)	radio advertisement	
Navaei et.al (2015), Mosavi Esmaeilzdeh (2015), Ashouri (2016), Hassanzadeh & Danaei (2016)	cell phone advertisement	
Navaei et.al (2015), Hosseinzadeh et.al (2016), Mirzadeh & Mehrmanesh (2016)	creating TV advertisement	
Ahmadi (2013), Salimi Fard & Pourdehghan (2016), Hoang et.al (2013)	attraction of rewards	
siadat & Sakhi Jourashk (2014), Koumra et.al (2016)	being benevolent	

Darzaian (2016), Zabkouich & Louksten (2011), Abdolsalam et.al (2016)	coordination in speaking and behavior	
Ahmadi (2013), Salimi Fard & Pourdehghan(2016), Honari et.al (2015)	reality of messages	
Madhoushi (2013), Salimi Fard & Pourdehghan (2016), Houang et.al (2013)	graphic-designing	
Shahmohammadi (2011), Madhoushi (2013), Salimi Fard & Pourdehghan(2016)	psychological – geographical division	
Mirzadeh & Mehrmanesh (2016), Zabkouich & Lokesten (2011), Menroz et.al (2012), Choubhai et.al (2013), Mokan & Oltz (2014), Ayono et.al (2014)	advertisement on web	
Dehdashti Shahrokh (2009), Shahmohammadi (2011), Mirzadeh & Mehrmanesh (2016), Menroz et.al (2012), Choubhai et.al (2013), Mokan & Oltz (2014), Ayono et.al (2014), Henri et.al (2015), Yazghiro- Basteh (2015), Lin (2016)	internet & cell phone advertisement	
Shahmohammadi (2011), Sanei et.al (2016), Menroz et.al (2012), Chobai et.al (2013), Ayono et.al (2014), Henri et.al (2015)	selecting advertising media	
Shahmohammadi (2011), Sanei et.al (2016), Howang et.al (2013), Henri et.al (2015), Yazghirdo-Yasteh (2015), Martina Santa et.al (2017)	artistic aspect of advertisement	
Kashizadeh (2010), Hoshivar(2014), Pelpin et.al (2017)	gender of speaker in the advertisement	
Hosseinzadeh et.al (2016), oribeh (2016), Madra et.al (2017)	text repetition	
Shahmohammadi (2011), Hassanzadeh & Danaei (2016), Abri & Joumar (2014), Ayouno et.al (2014), Oribeh (2016), Pachman (2016), Bak (2016), Madra et.al (2017)	advertisement	
Samsam Shariat (2007), Sohrabi Ranani (2016), Daneshvar et.al (2016), Bak (2016), Kiji-Kali (2017)	factors relating to usage	environment

Fakhari et.al (2012), Mirzadeh & Mehrmanesh (2016), Howang et.al (2013), Bak (2016), kiji-Kali (2017)	factors relating to facilitation	
Dargi & Mohammadian (2010), Hamidizadeh (2013), Darzban (2016), Belman et.al (2012)	awareness of different status of the environment status	
Dargi & Mohammadian (2010), Mirzadeh & Mehrmanesh (2016)	awareness of access location of objective group to the product	
Dargi & Mohammadian (2010), Salimifard & Pourdehghan (2016)	awareness of norms	
Dargi & Mohammadian (2010), Hamidizadeh (2013), Darzban (2016), Belman et.al (2012), Koumra et.al (2016)	awareness of religions and religious values	
Dargi & Mohammadian (2010), Hassanzadeh & Danaei (2016)	awareness of method of evaluating services in relation to other services	
Hassazadeh & Danaei (2016), Radmiker (2015)	attention to security	
Samsam Shariat (2007), Mirzadeh & Mehrmanesh (2016), Belman et.al (2012)	economic beliefs	
Salimifard & Pourdehghan (2016), Pilay et.al (2012), Howang et.al (2013), Komra et.al (2016)	ethical beliefs	
Salimifard & Pourdehghan (2016)	Islamic teaching	
Samsam Shariat (2007), Daneshvar et.al (2016), Radmiker (2015)	strategic look to advertisement plan	
Dargi & Mohamadian (2010), Ansari & Rasei (2016), Abri & Joumar (2014)	market share	
Dargi & Mohamadian (2010), Darzain (2016)	obviating primary needs of audiences	

Salimi Fard & Pourdehghan (2016), Kim et.al (2012), Howang et.al (2013)	reference groups	
Sohrabi Ranin (2016), Fontini et.al (2009), Belman et.al (2012)	awareness of services	factors relating to services
Salimi Fard & Pourdehghan (2016), Abri & Joumar (2014), Radmiker (2015)	stating potential benefits of purchase ad services consumption	
Dargi & Mohamadian (2010), Hosseinzadeh et.al (2016), Fontini et.al (2009)	rate of services complexity	
Samsam Shariat (2007), Darzaian (2016), Fontini et.al (2009), Radmiker (2015)	display of expected experience	
Khashizadeh (2010), Darzain (2016)	rate of familiarity of consumer with services	
Ansari & Rasaei (2016), Belman et.al (2012)	market researches	
Hosseinzadeh et.al (2016), Mirzadeh & Mehrmanesh (2016)	record of exploiting services	
Dargi & Mohammadian (2010), Ansari & Rasaei (2016), Lousi (2012), Park (2015), Marion et.al (2017)	being unique	
Dehdashti Shahrokh (2009), Daneshvar et.al (2016), Radmiker (2015), Jankoski et.al (2016)	price, benefits	
Daneshvar et.al (2016), Fontini et.al (2009), Jankoski et.al (2016)	relation with customer	
Samsam Shariat (2007), Daneshvar et.al (201), Fontini et.al (2009), Jankoski (2016)	supporting customer	
Dehdashti Shahrokh (2009), Ansari & Rasaei (2016), Fontini et.al (2009), Radmiker (2015), Jankoski (2016)	feedback	

2- Research plan

The present study is of descriptive applied type which has been performed with survey method. The statistical population of this study includes all managers and experts of Eghtesad

Novin Bank. The statistical population of the present study contains all people who are exposed to advertisements of Eghtesad Novin Bank across the country. For selecting samples, Cochran formula is used for unlimited societies that based on this formula, the samples number were determined 400 persons for each province. The sample people were selected based on the following sampling method. At first, using stratified sampling method , the samples were randomly selected from five geographical regions of the country (north, south, east, west and center) of nine provinces (Mazandaran, Khorasan Razavi, Zanjan, Chahar Mahal Bakhtiari , Kerman, East Azarbaijan , Kermanshah, Ilam and Tehran).

3- Variables and research model

In this study, fuzzy Delphi technique has been used for identifying main indicators. For ranking the final criteria and sub standards, multi-criteria decision making models and AHP model have been used with fuzzy approach. For analysis of obtained data, Excel software and coding in the environment of this software have been used. The results obtained from data analysis have been provided in the present chapter.

3-1 Identifying criteria and sub standards of factors effective on advertisement effectiveness

At the first step, the primary indicators were identified and selected. In the first stage, by reviewing the literature and performed studies, a large number of indicators were extracted from the literature. For screening the indicators and identifying the final indicators, fuzzy Delphi approach has been used. The expert views have been collected in the field of rate of significance of main indicators. Fuzzy average and defuzzification output of rates relating to indicators are presented in table 2. The rate of defuzzification greater than 7 is acceptable and every indicator which has the score higher than 7 is confirmed (Wu& Fang, 2011).

Table 2- fuzzy average and fuzzy screening of indicators of factors effective on advertisement effectiveness

Result	Crisp	mean	U	M	L	
acceptance	7.15	(2, 7.97, 9)	9	7.97	2	Q01
acceptance	7.84	(6, 8.01, 9)	9	8.01	6	Q02

Result	Crisp	mean	U	M	L	
acceptance	7.92	(6, 8.13, 9)	9	8.13	6	Q03
acceptance	7.39	(4, 7.83, 9)	9	7.83	4	Q04
acceptance	7.70	(6, 7.8, 9)	9	7.80	6	Q05
rejection	6.68	(2, 7.28, 9)	9	7.28	2	Q06
acceptance	7.71	(6, 7.82, 9)	9	7.82	6	Q07
acceptance	7.66	(6, 7.75, 9)	9	7.75	6	Q08
acceptance	7.02	(2, 7.78, 9)	9	7.78	2	Q09
acceptance	7.27	(2, 8.15, 9)	9	8.15	2	Q10
acceptance	7.09	(2, 7.88, 9)	9	7.88	2	Q11
rejection	6.85	(2, 7.52, 9)	9	7.52	2	Q12
acceptance	7.43	(4, 7.9, 9)	9	7.90	4	Q13
rejection	6.71	(3, 7.07, 9)	9	7.07	3	Q14
acceptance	7.17	(2, 8.01, 9)	9	8.01	2	Q15
acceptance	7.13	(2, 7.95, 9)	9	7.95	2	Q16
rejection	6.85	(4, 7.02, 9)	9	7.02	4	Q17
acceptance	7.43	(4, 7.89, 9)	9	7.89	4	Q18
acceptance	7.70	(6, 7.8, 9)	9	7.80	6	Q19
rejection	6.80	(3, 7.21, 9)	9	7.21	3	Q20
acceptance	7.75	(6, 7.88, 9)	9	7.88	6	Q21
rejection	6.62	(3, 6.93, 9)	9	6.93	3	Q22
acceptance	7.61	(6, 7.66, 9)	9	7.66	6	Q23
acceptance	7.22	(2, 8.08, 9)	9	8.08	2	Q24
acceptance	7.04	(2, 7.81, 9)	9	7.81	2	Q25
acceptance	7.71	(6, 7.82, 9)	9	7.82	6	Q26
rejection	6.65	(3, 6.97, 9)	9	6.97	3	Q27
acceptance	7.98	(6, 8.22, 9)	9	8.22	6	Q28
acceptance	7.70	(6, 7.79, 9)	9	7.79	6	Q29
rejection	6.54	(3, 6.81, 9)	9	6.81	3	Q30
acceptance	7.19	(4, 7.53, 9)	9	7.53	4	Q31
acceptance	7.79	(5, 8.18, 9)	9	8.18	5	Q32

Result	Crisp	mean	U	M	L	
acceptance	7.53	(4, 8.05, 9)	9	8.05	4	Q33
acceptance	7.05	(2, 7.82, 9)	9	7.82	2	Q34
acceptance	7.43	(4, 7.89, 9)	9	7.89	4	Q35
rejection	6.92	(3, 7.38, 9)	9	7.38	3	Q36
acceptance	7.02	(2, 7.78, 9)	9	7.78	2	Q37
rejection	6.91	(3, 7.37, 9)	9	7.37	3	Q38
rejection	6.96	(2, 7.68, 9)	9	7.68	2	Q39
rejection	6.99	(2, 7.74, 9)	9	7.74	2	Q40
acceptance	7.42	(4, 7.88, 9)	9	7.88	4	Q41
acceptance	7.10	(2, 7.9, 9)	9	7.90	2	Q42
rejection	6.59	(2, 7.13, 9)	9	7.13	2	Q43
rejection	6.55	(2, 7.08, 9)	9	7.08	2	Q44
acceptance	7.77	(6, 7.9, 9)	9	7.90	6	Q45
rejection	6.91	(2, 7.62, 9)	9	7.62	2	Q46
acceptance	7.09	(2, 7.88, 9)	9	7.88	2	Q47
acceptance	7.68	(6, 7.78, 9)	9	7.78	6	Q48
acceptance	8.19	(6, 8.53, 9)	9	8.53	6	Q49
rejection	6.88	(2, 7.57, 9)	9	7.57	2	Q50
acceptance	7.75	(5, 8.12, 9)	9	8.12	5	Q51
acceptance	7.03	(2, 7.8, 9)	9	7.80	2	Q52
rejection	6.96	(4, 7.19, 9)	9	7.19	4	Q53
acceptance	7.04	(2, 7.8, 9)	9	7.80	2	Q54
rejection	6.96	(3, 7.45, 9)	9	7.45	3	Q55
acceptance	7.02	(2, 7.78, 9)	9	7.78	2	Q56
acceptance	7.92	(5, 8.38, 9)	9	8.38	5	Q57
acceptance	7.00	(2, 7.76, 9)	9	7.76	2	Q58
rejection	6.90	(1, 7.86, 9)	9	7.86	1	Q59
acceptance	7.10	(4, 7.4, 9)	9	7.40	4	Q60
acceptance	7.86	(6, 8.04, 9)	9	8.04	6	Q61
rejection	6.97	(2, 7.71, 9)	9	7.71	2	Q62

Result	Crisp	mean	U	M	L	
acceptance	7.02	(2, 7.79, 9)	9	7.79	2	Q63
acceptance	7.73	(6, 7.84, 9)	9	7.84	6	Q64
acceptance	7.01	(2, 7.76, 9)	9	7.76	2	Q65
rejection	6.65	(3, 6.98, 9)	9	6.98	3	Q66
acceptance	7.01	(2, 7.77, 9)	9	7.77	2	Q67
acceptance	7.90	(6, 8.11, 9)	9	8.11	6	Q68
acceptance	7.55	(5, 7.83, 9)	9	7.83	5	Q69
acceptance	8.02	(6, 8.28, 9)	9	8.28	6	Q70
acceptance	7.95	(6, 8.17, 9)	9	8.17	6	Q71
acceptance	7.03	(2, 7.79, 9)	9	7.79	2	Q72
acceptance	7.65	(5, 7.98, 9)	9	7.98	5	Q73
acceptance	7.00	(1, 8, 9)	9	8.00	1	Q74
acceptance	7.04	(2, 7.8, 9)	9	7.80	2	Q75
acceptance	7.03	(2, 7.8, 9)	9	7.80	2	Q76
acceptance	7.30	(4, 7.7, 9)	9	7.70	4	Q77
rejection	6.88	(2, 7.57, 9)	9	7.57	2	Q78
acceptance	7.63	(4, 8.19, 9)	9	8.19	4	Q79
acceptance	7.65	(5, 7.98, 9)	9	7.98	5	Q80
rejection	6.59	(3, 6.88, 9)	9	6.88	3	Q81
acceptance	7.09	(2, 7.88, 9)	9	7.88	2	Q82
rejection	6.83	(2, 7.5, 9)	9	7.50	2	Q83
acceptance	7.12	(2, 7.93, 9)	9	7.93	2	Q84

From the 84 indicators which were assessed, 24 indicators which have acquired the average score less than 7 are omitted. So, 60 indicators are finally confirmed (Adelazaret.al, 1995).

The remained indicators will be finally examined in the format of a 60-question questionnaire with analytical hierarchy process.

3-2 determining the priority of factors effective on effectiveness of advertisement using fuzzy AHP technique

For determining the priority of major criteria, fuzzy analytical hierarchy (FAHP) has been used. The analysis process is as follows.

1- Dual comparison of major criteria based on objective and determining the weight of major criteria

2- Dual comparison of sub standards of each criterion and determining sub standards of each cluster

3- Multiplication of the weight of sub standards in the weight of respective criterion and determining the final weight of sub standards

For dual comparison of elements, 9-degree Saaty scale has been used. 9-degree Saaty scale has been provided by Thomas Saaty , codifier of analytical hierarchy theory. Also, in this study fuzzy approach has been used for making the values quantitative. So, Saaty fuzzy range has been used.

Table 3- fuzzy range equal to 9-degree Saaty scale in AHP technique. Habibi et.al , 2014, 77

reversed fuzzy equivalent	fuzzy equivalent	the word term of comparison status of i to j
(1,1,1)	(1, 1, 1)	Preferred Equally
$\left(\frac{1}{3},\frac{1}{2},1\right)$	(1, 2, 3)	intermediate
$\left(\frac{1}{4},\frac{1}{3},\frac{1}{2}\right)$	(2, 3, 4)	Preferred moderately
$\left(\frac{1}{5},\frac{1}{4},\frac{1}{3}\right)$	(3, 4, 5)	intermediate
$\left(\frac{1}{6},\frac{1}{5},\frac{1}{4}\right)$	(4, 5, 6)	Preferred Strongly
$\left(\frac{1}{7},\frac{1}{6},\frac{1}{5}\right)$	(5, 6, 7)	intermediate
$\left(\frac{1}{8},\frac{1}{7},\frac{1}{6}\right)$	(6, 7, 8)	very strongly Preferred
$\left(\frac{1}{9},\frac{1}{8},\frac{1}{7}\right)$	(7, 8, 9)	intermediate
$\left(\frac{1}{9},\frac{1}{9},\frac{1}{9}\right)$	(9, 9, 9)	Extremely Preferred

(source in papersL Khorshid & Zabihi, Lee my H.I; Wen-Chin Chen, Ching-Jan, Chang, 2008, 101)

The pattern of criteria and sub standards has been drawn in the figure using AHP technique. Also, the criteria and sub standards of the study have been named with numeric index so that the research flow is easily detectable.

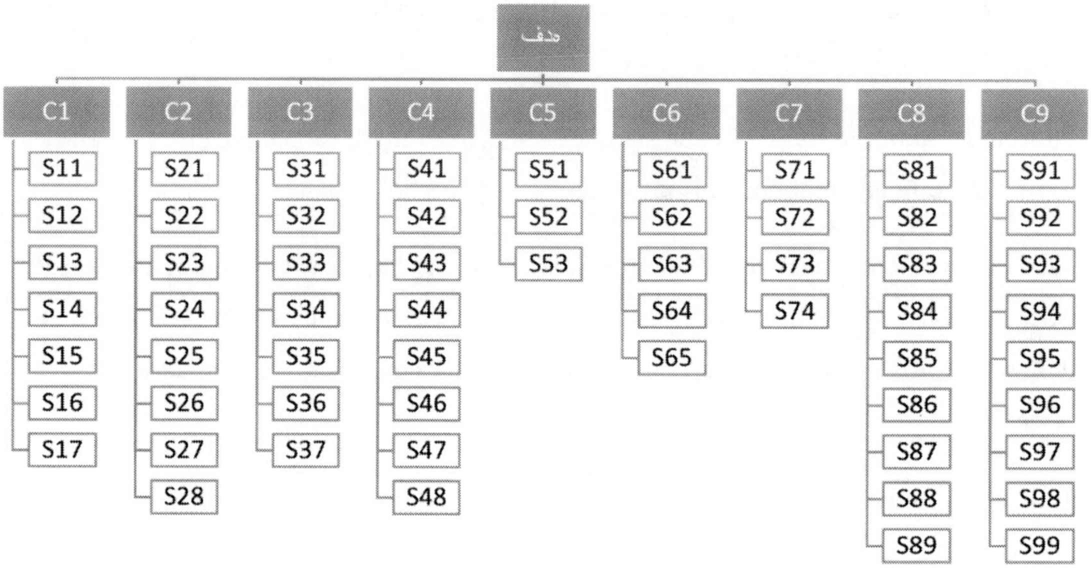

Figure 1- hierarchical model of criteria and sub standards of factor effective on advertisement effectiveness

1- Determining the priority of major criteria based on objective

In the first step, the major criteria based on the objective have been compared dually. Dual comparison is very simple and all elements of each cluster should be compared two by two. Therefore, if there is n element in a cluster, $\frac{n(n-1)}{2}$ comparisons will be performed. Since there are four criteria, so the number of performed comparisons is equal to:

$$\frac{n(n-1)}{2} = \frac{9(9-1)}{2} = 36$$

So, 36 dual comparisons have been performed from group perspective of experts. The experts view has been made quantitative using fuzzy scale. Collecting experts view has been collected with 9-degree Saaty range. So, 6 dual scales of four major four criteria have been performed based on the experts view. Using geometric average the experts views have been aggregated. Fuzzy aggregation will be calculated as follows:

Equation 1

$$F_{AGR} = (\prod(l), \prod(m), \prod(u))$$

After matrix formation of the obtained dual comparisons, the special vector has been calculated. At first, the fuzzy extension of each row is calculated. Every entry of dual comparison matrix $\tilde{X}$

Is displayed as $\tilde{x}_{ij}$.

The fuzzy extension of every row is also displayed with symbol $\tilde{S}_i$. Therefore, the fuzzy extension of each row is calculated as follows:

Equation 2

$$\tilde{S}_i = \sum_{j=1}^{n} x_{ij}$$

Then, the fuzzy extension of every row elements will be as follows:

Fuzzy extension of row 1 (1,1,1) ⊕ (1,27,1,47,1,75) ⊕ (0.47, 0.55, 0.66) ⊕ (1.52, 1.81, 2.14) ⊕ (0.93, 1.14, 1.39) ⊕ (1.26, 1.61, 2.04) ⊕(1.25, 1.51, 1.84) ⊕(0.78, 0.97, 1.24) ⊕ (0.59, 0.75, 0.96) ⊕ (9.08, 10.8, 13.02)

Fuzzy extension of row 2 (0.57, 0.68, 0.79)⊕(1, 1, 1)⊕(0.58, 0.7, 0.89)⊕(1.84, 2.28, 2.75)⊕(0.68, 0.82, 1.02)⊕(1.42, 1.81, 2.34)⊕(1.64, 2.04, 2.53)⊕(0.71, 0.88, 1.12)⊕(0.73, fuzzy extension of row 3 (1.52, 1.82, 2.11)⊕(1.13,)0.9, 1.1)⊕ = (9.17, 11.11, 13.55 1.42, 1.72)⊕(1, 1, 1)⊕(1.89, 2.34, 2.88)⊕(0.7, 0.86, 1.09)⊕(1.45, 1.78, 2.22)⊕(1.55, 1.88, 2.29)⊕(0.57, 0.71, 0.93)⊕(0.67, 0.8, 0.96)⊕ = (10.47, 12.61, 15.2)

Fuzzy extension of row 4 (0.47, 0.55, 0.66)⊕(0.36, 0.44, 0.54)⊕(0.35, 0.43, 0.53)⊕(1, 1, 1)⊕(0.97, 1.21, 1.51)⊕(0.96, 1.33, 1.8)⊕(2.14, 2.7, 3.37)⊕(0.94, 1.21, 1.56)⊕(1.3, 1.69, 2.16)⊕ = (8.49, 10.57, 13.14)

Fuzzy extension of row 5 (0.72, 0.88, 1.07) ⊕ (0.98, 1.22, 1.48)⊕(0.92, 1.17, 1.43)⊕(0.66, 0.82, 1.03)⊕(1, 1, 1)⊕(0.67, 0.83, 1.04)⊕(0.23, 0.29, 0.36)⊕(0.28, 0.32, 0.39)⊕(0.3, 0.38, 0.48)⊕ = (5.76, 6.91, 8.29)

Fuzzy extension of row 6 (0.49, 0.62, 0.79)⊕(0.43, 0.55, 0.7)⊕(0.45, 0.56, 0.69)⊕(0.55, 0.75, 1.04)⊕(0.96, 1.2, 1.48)⊕(1, 1, 1)⊕(0.34, 0.41, 0.51)⊕(0.27, 0.34, 0.44)⊕(0.24, 0.31, 0.41)⊕ = (4.74, 5.75, 7.08)

Fuzzy extension of row 7 (0.54, 0.66, 0.8)⊕(0.4, 0.49, 0.61)⊕(0.44, 0.53, 0.64)⊕(0.3, 0.37, 0.47)⊕(2.81, 3.5, 4.3)⊕(1.95, 2.43, 2.96)⊕(1, 1, 1)⊕(0.34, 0.42, 0.54)⊕(0.29, 0.36, 0.47)⊕ = (8.06, 9.76, 11.78)

Fuzzy extension of row 8 (0.8, 1.03, 1.28)⊕(0.89, 1.14, 1.41)⊕(1.08, 1.41, 1.76)⊕(0.64, 0.82, 1.06)⊕(2.55, 3.1, 3.63)⊕(2.25, 2.97, 3.72)⊕(1.86, 2.39, 2.94)⊕(1, 1, 1)⊕(0.58, 0.69, 0.83)⊕ = (11.66, 14.56, 17.64)

Fuzzy extension of row 9 (1.04, 1.33, 1.69)⊕(0.91, 1.11, 1.37)⊕(1.04, 1.25, 1.5)⊕(0.46, 0.59, 0.77)⊕(2.08, 2.63, 3.3)⊕(2.42, 3.23, 4.1)⊕(2.15, 2.74, 3.43)⊕(1.2, 1.46, 1.72)⊕(1, 1,)1)⊕ = (12.29, 15.35, 18.88

Then, the fuzzy sum of total priorities column elements is calculated

Equation 3

$$\sum \tilde{S}_i = \sum_{i=1}^{n} \sum_{j=1}^{n} x_{ij}$$

The sum of elements of the priorities column of major criteria will be as follows:

$$\sum_{i=1}^{4} \sum_{j=1}^{4} M_g^j = (79.72, 97.41, 118.59)$$

For normalization of each criterion priorities, that criterion total rates sum should be divided into sum of all priorities (column elements), since the rates are fuzzy, so fuzzy sum of every row is multiplied in summation reverse. The summation reverse should be calculated.

Equation 4

$$if\ \tilde{F} = (l, m, u)\ then\ \tilde{F}^{-1} = \left(\frac{1}{u}, \frac{1}{m}, \frac{1}{l}\right)$$

Then, based on the equation 4, we will have

$$\left(\sum_{i=1}^{n} \sum_{j=1}^{n} M_g^j\right)^{-1} = (0.01, 0.01, 0.01)$$

$$P_k = \sum_{i=1}^{n} M * \left(\sum_{i=1}^{n} \sum_{j=1}^{n} M_g^j\right)^{-1}$$

So, the results obtained from normalization of obtained rates will be as follows:

$\widetilde{W}_{C1} =$ (0.08, 0.11, 0.16)

$\widetilde{W}_{C2} =$ (0.08, 0.11, 0.17)

$\widetilde{W}_{C3} =$ (0.09, 0.13, 0.19)

$\widetilde{W}_{C4} =$ (0.07, 0.11, 0.16)

$\widetilde{W}_{C5} =$ (0.05, 0.07, 0.1)

$\widetilde{W}_{C6}$ = (0.04, 0.06, 0.09)

$\widetilde{W}_{C7}$ = (0.07, 0.1, 0.15)

$\widetilde{W}_{C8}$ = (0.1, 0.15, 0.22)

$\widetilde{W}_{C9}$ = (0.1, 0.16, 0.24)

Each of the obtained rates is fuzzy and normalized weight relating to major criteria. There are various methods for defuzzifcation of the obtained rates. In this study, Bojadziev proposed method has been used for defuzzification.

Table 4- defuzzification and computation of final and normal weight of criteria of factors effective on advertisement effectiveness

Normal	Deffuzy	X3max	X2max	X1max	Crisp
0/111	0/117	0/113	0/115	0/117	C1
0/114	0/120	0/115	0/118	0/120	C2
0/103	0/137	0/133	0/135	0/137	C3
0/108	0/113	0/112	0/113	0/113	C4
0/070	0/073	0/072	0/073	0/073	C5
0/060	0/063	0/062	0/063	0/063	C6
0.101	0/107	0/103	0/105	0/107	C7
0.149	0/157	0/153	0/155	0/157	C8
0.158	0/167	0/163	0/165	0/167	C9

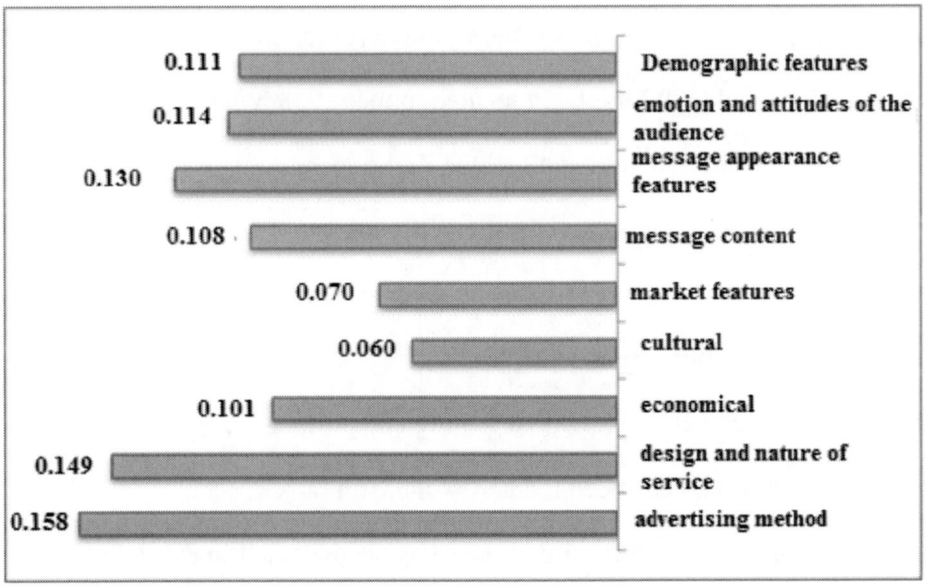

Figure 2- graphical display of significance of criteria of factors effective on advertisement effectiveness

Based on the obtained special vector:

- Demographic features with weight 0.111 have been located in rank five

- Individual feelings and attitude of the addressee with weight 0.114 has been located in rank four.

- The message surface features with weight 0.13 has been located in rank three.

- The message content with weight 0.108 has been located in rank six

- The market feature with weight 0.07 has been located in rank eight.

- Cultural with weight 0.06 has been located in rank nine.

- Economic with weight 0.10 has been located in rank seven.

- Plan and nature of services with weight 0.149 has been located in rank second.

- Advertisement methods with weight 0.158 has been located in rank first.

The inconsistency rate of performed comparisons has been obtained 0.030 which is smaller than 0.1 and so the performed comparisons is sufficed.

2- Comparison and determining priority of sub standards

At first, the experts' views have been collected in the dual comparison of sub standards. Then, the performed calculations for fuzzification of the experts views average have been provided for determining the priority of sub standards.

Priority of sub standards of demographic features

The audience job with weight 0.151 has been located in the third rank.

The audience income with weight 0.093 has been located in the seventh rank.

The audience religion with weight 0.16 has been located in the second rank.

Social class with weight 0.141 has been located in the fifth rank.

Awareness of gender differences with weight 0.173 has been located in the first rank.

The individual knowledge level with weight 0.135 has been located in the sixth rank.

Priority of sub standards of the audience individual feeling and attitude

Positive mentality with weight 0.124 has been located in the fifth rank.

Feeling of honesty in advertisements with weight 0.089 has been located in the eighth rank.

Threatening feeling with weight 0.14 has been located in the third rank.

Mental image with weight 0.102 has been located in the seventh rank.

Attitude towards brand with weight 0.121 has been located in the sixth rank.

Convincing audiences with weight 0.149 has been located in the first rank.

Attitude towards advertisement with weight 0.143 has been located in the second rank.

The individual perception of accuracy of advertisement with weight 0.133 has been located in the fourth rank.

Priority of sub standards of the message surface features

Then, based on the results of performed dual comparisons:

Beautiful and appropriate combination of colors with weight 0.143 has been located in the fourth rank.

Attention to reduction of the user waiting time with weight 0.159 has been located in the second rank.

Graphics- design with weight 0.131 has been located in the sixth rank.

Creativity in the message with weight 0.137 has been located in the fifth rank.

Emotional music with weight 0.099 has been located in the seventh rank.

Text repetition with weight 0.153 has been located in the third rank.

Access to required and sufficient information with weight 0.178 has been located in the first rank.

Determining the priority of sub standards of the message content

Benevolent with weight 0.093 has been located in the eighth rank.

Coordination in speech and behavior with weight 0.125 has been located in the fourth rank.

Reality of the messages with weight 0.119 has been located in the sixth rank.

Using women and children images with weight 0.131 has been located in the third rank.

Gender of speaker in the advertisement with weight 0.106 has been located in the seventh rank.

Using trustful and famous people with weight 0.16 has been located in the first rank.

Loveliness of famous people for the observer with weight 0.122 has been located in the fifth rank.

Familiarity of the famous person with the observer with weight 0.144 has been located in the second rank.

Priority of sub standards of the market features

Then, based on the results of the performed dual comparisons:

Factors relating to its application in the market with weight 0.326 have been located in the second rank.

Awareness of different aspects of status and environment with weight 0.283 has been located in the third rank.

Awareness of evaluation method of service receiver with weight 0.391 has been located in the first rank.

Priority of cultural sub standards

Then, based on the results of performed dual comparisons:

Awareness of norms with weight 0.199 has been located in the third rank.

Awareness of religion and religious values with weight 0.238 has been located in the first rank.

Ethical beliefs with weight 0.18 have been located in the fourth rank.

Obviating primary needs of the audiences with weight 0.164 has been located in the fifth rank.

Reference groups with weight 0.219 have been located in the second rank.

Determining the priority of economic sub standards

Then, based on the results of the performed dual comparison:

Economic beliefs with weight 0.261 have been located in the second rank.

Strategic look at an advertising program with weight 0.215 has been located in the fourth rank.

Market share with weight 0.274 has been located in the first rank.

Attention to economic security with weight 0.251 has been located in the third rank.

Priority of the plan and services nature sub standard

Awareness of services with weight 0.104 has been located in seventh rank.

Expressing potential benefits of purchase and services consumption with weight 0.117 has been located in the third rank.

Services complexity rate with weight 0.089 has been located in the ninth rank.

Uniqueness with weight 0.108 has been located in the sixth rank.

Rate of familiarity of the services consumer with weight 0.12 has been located in the second rank.

Price and allowances with weight 0.114 has been located in the fourth rank.

Relation with the customer with weight 0.111 has been located in the fifth rank.

Displaying the expected experience with weight 0.098 has been located in the eighth rank.

Supporting customers with weight 0.139 has been located in the first rank.

Priority of sub standards of the advertising methods

Beginning the advertisement with a challenge with weight 0.079 has been located in the ninth rank.

Beginning an advertisement with a statement and explanation with weight 0.107 has been located in the fifth rank.

Using stimuli with diversions more than usual and expected with weight 0.097 has been located in the seventh rank.

The advertisement message text being predicating with weight 0.126 has been located in the third rank.

Using diverse methods with weight 0.088 has been located in the eighth rank.

Media advertisement with weight 0.113 has been located in the fourth rank.

Dimensions of TV advertisement with weight 0.151 have been located in the first rank.

Psychological, geographical division with weight 0.107 has been located in the fifth rank.

Advertising media selection with weight 0.132 has been located in the second rank.

Determining the final priority of major criteria with FAHP technique

For determining final priority of major criteria using FAHP technique, the weights relating to major criteria (W1) and weight of indicators based on each criteria (W2) are needed. The results of comparison of the research sub criteria relating to them forms matrix W2. For determining the final priority of indicators with AHP technique, it is sufficient that the weight of indicators based on every criterion (W2) is multiplied in the major criteria (W1). Any of these matrices has been calculated in the previous steps. Based on the obtained output, it is obvious that the market share item, one of economic factors with weight 0. 277 is the most important indicator in the existing indicators. Item of awareness of evaluation method of service receiver in relation to other services with weight of 0.0273 has the second priority. Item of economic believes, one of economic factors, has the third priority.

So, the research model is as follows:

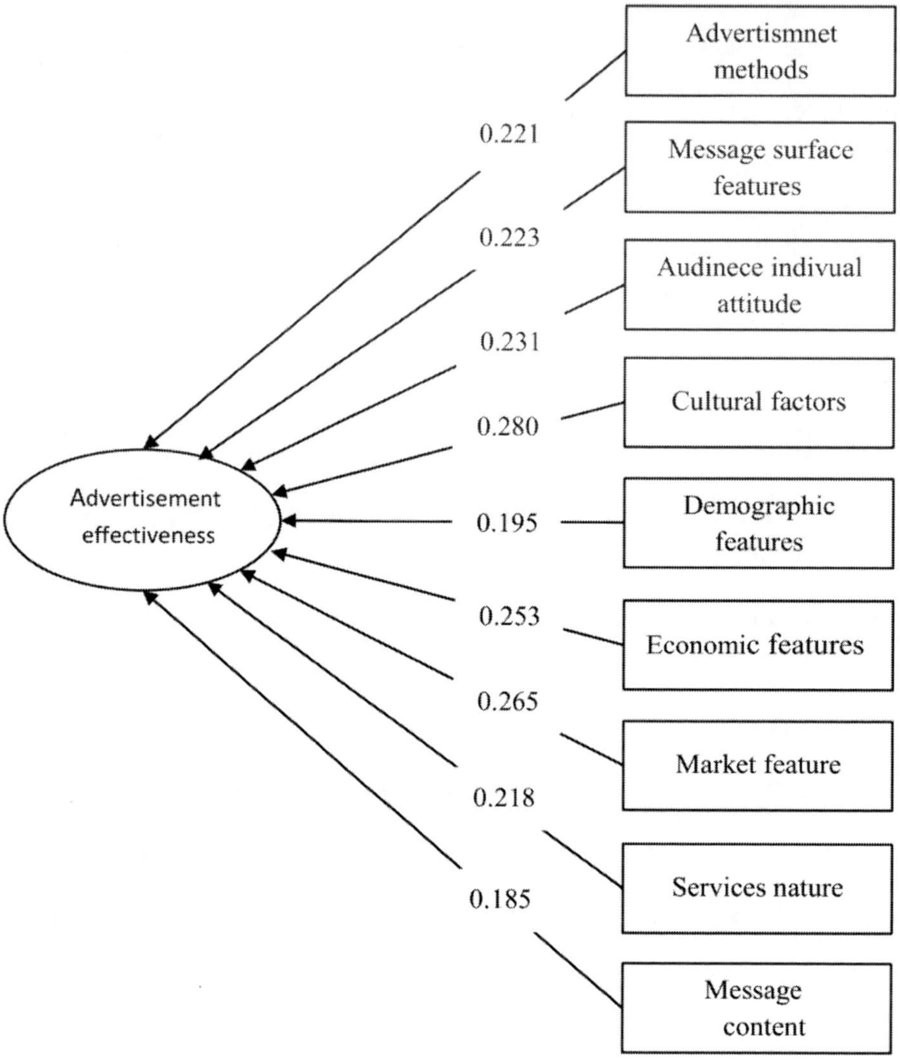

4- Conclusion:

In this study, identification and ranking of factors effective on designing the native model of advertisement in Eghtesad Novin bank has been addressed. In this first step, the primary indicators were identified and selected. In the first stage, by reviewing the literature and performed studies, a great number of indicators were extracted from the literature. For screening the indicators and identifying final indicators, fuzzy Delphi approach has been used. The experts' views have been collected in the field of significant rate of major indicators. From 84 indicators which were assessed, 24 indicators which have acquired score less than 7 are omitted. So, 60 indicators were finally confirmed. Then, regarding the performed calculations, the final weight of any of the model indicators have been calculated with fuzzy AHP. Based on the obtained output, it is obvious that the market share of economic factors with weight of 0.027 is the most important indicator in the existing indicators. The item of awareness of evaluating services in relation to other services is one of market features factors with weight 0.0273 which has second priority. The item of economic beliefs with weight 0.0264 is in the third priority.

Today, considering extension of competition in the international level and formation of world competition, all countries of the world may be successful in this arena to coordinate them with this trend and create more value for their customers by services and products they provide. In the today modern world, almost all people are exposed to advertisements and other communicational activities. Effectiveness of advertisement is one of important issues in the advertisement domain. In this regard, Mansouri Moyed (2005) showed that attention attraction, commercial advertisements of Maskan Bank are effective but in respect of other variables, that is, creating interest, stimulation of tendency and measurement to purchase, commercial advertisements of Maskan Bank are not effective. Samsam Shariat (2007) showed that TV channels and cartoon animations , attractive package, colors relating to the product and innovation, noticing safety feeling and associated symbols, informing of quality, background and price, dialects and common phrases, diversity in the message content, in TV, broadcasting from 18-23 and before beginning programs (in media) have been effective factors in effectiveness of commercial effectiveness from people viewpoint. Mosavi Esmaeilzadeh (2015) showed that oral advertisements (in 4 dimensions) and after that TV advertisement (just in 2 dimensions) are effective. Also, radio, internet and billboard advertisement are not effective. And also Friedman test was used for ranking advertisement tools. Salimi Fard & Pourdehghan (2016) showed that among factors effective on attitude to cell phone advertisement, components of general attitude towards advertisement, knowledge level and value have respectively the greatest significance and components of time of presentation and reference groups with the least significance in effecting the individuals attitude towards cell phone advertisement. Ansari & Rasaei (2016) showed that the creativity of advertisement message , market research, competition, market share, uniqueness and relation with customer are effective on the effectiveness of brand advertisement. Louis (2012) showed that advertising in the urban public spaces will have a great impact in the consumers' mentality. Since some advertisements are transitory and are often displayed in a short period of time and just in several spaces, but advertisement in public spaces exist everywhere and is similar to advertising through print and media, but since they have lower cost for production and distribution, have lower cost in this respect. Zabckowich & Loukston (2011) showed that in necktie advertisements use more of sensory requests and information signs. Findings showed that in western and European countries, people establish more contact due to globalization increase. They declared that content of printed advertisements, includes visual and text advertisements which are effective in establishing relation with many audiences. Also, ethnic identity as an important cultural variable increases attention to consumer purchase behavior. Besides that, they found out that the image of people, models, ceremonies and managers constitute an important part of images and background image plays an important role in effectiveness of visual factors in advertisement. Mac Ki- Nesbiat et.al (2011) showed that younger adults remind emotional messages specially negative ones better than others and older adults more prefer positive and logical messages to negative and emotional ones, but attitude towards advertisement among younger adults is not different. Also, they found that older adults have the most positive attitude towards advertisements.

Perzouti (2015) showed that advertising models are different in various ages and each age is consistent with a special model and using various and consistent models with people conditions may make the advertisement more effective. Abri & Joumar (2014) showed that advertising messages are effective on the consumer behavior and the advertiser may use various media for communicating consumer and media schedule is effective on the advertisement success. Attention to environmental factors where the customer lives are important factors in the advertisement success. Madra et.al (2017) showed that the advertisement text, text repetition impact the advertisement effectiveness. Maryion et.al (2017) showed that simultaneous use of several media like TV and internet may increase effectiveness of the advertisement and considering two variables of gender and multimedia create a better individual attitude towards the message effectiveness.

In ranking the identified factors, advertisement methods have assigned the first rank among factors. It is suggested that in designing advertisement and selecting advertising instruments, the dimensions of TV advertising, selecting median and predicative message text should be noticed. Using advertising instrument like internet advertisement, adverting on public vehicles like buses and metro due to great number of people who use them can increase stability of the advertisements in the customers mind.

In codifying and designing bank advertisements, the design and services nature should be noticed. It is proposed that supporting customer, familiarity of consumer with services are considered and during the advertisement process, it is required that the potential benefits of purchase and services consumption are expressed.

In designing the advertisement and content of bank advertising message, it is required to use famous and trustful people for introducing the services. Also, children as future customers and audiences should be noticed since they are future potentials. Total advertisements provided by the bank through various media should move in a unit line so that the audience can establish a logical relation between the observed advertisements in various media that is total advertising efforts lead to synergy of advertisement effectiveness. This issue won't be acquired unless through a media planning and efficient advertisement.

In advertisements, the message surface features should be noticed. It is required that in advertisement, access to required and enough information for the customers, attention to reduction of expectation time of user for observing the advertisement and suitable repetition of the text are considered. Providing large information in an advertisement which has a short time and providing information that require special mathematical analysis should be avoided.

References:

Hosseinabadi, Maryam, Mahmoud Nematian, Mohammad Besharati, investigating the relation between mixed elements of marketing and customers satisfaction with Tehran

Sarmayeh bank branches, management and entrepreneurship studies, fall 2016, 2nd period, No. 3 (page 162-169)

Dargi Parviz, 2008, subjects and topics of marketing management with Iran market attitude, Tehran, Rasa publication

Sarmad, Zohreh, Bazargan Abbas, Hejazi, Elaheh, 2014, research methods in behavioral sciences, Tehran, Agah. 27-13

Salimi Fard, Khodakaram, Pourdehghan, Adel, 2016, identifying and ranking factors effective on attitude towards cell phone advertisement using AHP, Brand management, period 3, no. 6, summer 2016, page 133-160

Seyed Akbar Manjgani, Vahideh Tabasi Lotfabadi, 2015, factors effective on effectiveness of advertisement on financial stoop of legal clients of Mashhad City with AHP approach, finance journal, fall 2015, No. 75 (page 121-146)

Sanayei, Ali, Majid Mohammad Shafiei, Maedeh Amini Valashani, 2016, effectiveness of bank advertisement in customer attraction, commercial management searches, spring and summer 2016, No. 15 (page 185-210)

Mahmoudian, Mahmoud, Rabiei, Ali, Jamili, Bita. 2011, evaluation of effectiveness of Parsian bank advertisement and identifying the most important factors in increasing its effectiveness in Tehran city, modern marketing studies, summer 2011, No. 2 (page 17-40).

Manjagani, Seyed Akbar, Tabasi Lotfabadi, Vahideh, 2015, factors effective on advertisement effectiveness on financial stoop of legal clinets of Mashhad city with AHP approach, finance journal, fall 2015, No. 75, 1-121

Venous, Davar, Safaeian, Mitra, (2005), marketing bank services in Iran and bank services features, Tehran, Negah Danesh, 4th edition

Angulo, N. (2016). 10 digital marketing trends to watch in 2016 and beyond. Retrieved from: http://www.marketingdive.com/news/10-digital-marketing-trends-to-watch-in-2016-./and-beyond/411457

Bell Raoul, Buchner, Axel.(2018). Positive Effects of Disruptive Advertising on Consumer Preferences, Journal of Interactive Marketing 41 (2018) 1–13

Corvi. Elisabetta.(2010).The effectiveness of advertising: a literature review, 10th Global Conference on Business & Economics, ISBN : 978-0-9830452-1-2

Cramphorn Michael F.,(2011), Forum Gender effects in advertising, International Journal of .Market Research Society Vol. 53 Issue 2

Cronin, J. J. & Fox , G. L. (2010). "The implications of third party customer complaining for .advertising efforts". Journal of advertising. 39(2): 21-23

Decrop, A. (2007). The influence of message format on the effectiveness of print advertisements for tourism destinations. International Journal of Advertising, 26(4), 505–.525

Denis .A. Shakhov and ,Anna A. Panasenko.(2012).Evaluating Effectiveness of Bank Advertising in the Internet: Theory and Practice, World Applied Sciences Journal 18 (Special Issue of Economics): 83-90, 2012 ISSN 1818-4952

Denis 1,2 A. Shakhov and 3Anna A. Panasenko.(2012).Evaluating Effectiveness of Bank Advertising in the Internet: Theory and Practice, World Applied Sciences Journal 18 (Special Issue of Economics): 83-90, 2012

Durdana Ozretic-Dosen Ines Zizak , (2015),"Measuring the quality of banking services targeting student population", EuroMed Journal of Business, Vol. 10 Iss 1 pp. 98 - 117

Eckler, P., and Bolls, P. (2011). Spreading the Virus: Emotional Tone of Viral Advertising and Its Effect on Forwarding Intention and Attitudes. Journal of Interactive Advertising, .11(2): 1-11

Eric W.T. Ngai Ka-leung Karen Moon S.S. Lam Eric S. K. Chin Spencer S.C. Tao , (2015),"Social media models, technologies, and applications", Industrial Management & Data Systems, Vol. 115 Iss 5 pp. 769 – 802

Fotini Patsioura Maro Vlachopoulou Vicky Manthou, (2009),"A new advertising effectiveness model for corporate advertising web sites", Benchmarking: An International Journal, Vol. 16 Iss 3 pp. 372 - 386

Frederic Crane, , G. (2010). «Marketing FO Entrepreneurs: ConceptandApplication for new .Venture. New York: SAGE Pablication, Inc

Frolova. Svetlana.(2014).THE ROLE OF ADVERTISING IN PROMOTING A PRODUCT, Thesis CENTRIA UNIVERSITY OF APPLIED SCIENCES ,Degree Programme in Industrial Management May 2014, p"1-61

Henry Boateng Abednego Feehi Okoe , (2015),"Consumers' attitude towards social media advertising and their behavioural response", Journal of Research in Interactive Marketing, Vol. 9 Iss 4 pp. 299 -312

Henry Boateng Abednego Feehi Okoe , (2015),"Consumers' attitude towards social media advertising and their behavioural response", Journal of Research in Interactive Marketing, Vol. 9 Iss 4 pp. 299 -312

Huang. Jinsong, Song Sub, , Liuning Zhouc, , , Xi Liu .(2013).Attitude Toward the Viral Ad: Expanding Traditional Advertising Models to Interactive Advertising, Journal of Interactive Marketing, Volume 27, Issue 1, February 2013, Pages 36–46

Idris, I., M.S.A. Yajid and A. Khatibi, 2009. Personal and Psychological Factors-Does it .Impact the Choice of Advertising Medium? Journal of Social Sciences, 5(2): 104-111

Ioannou Myria, Nikolaos Boukas, Evi Skoufari .(2014).Examining the role of advertising on the behaviour of co-operative bank consumers, Journal of Co-operative Organization and Management, Volume 2, Issue 1, June 2014, Pages 24-33

Jankowski. Jarosław, Przemysław Kazienko, Jarosław Wątróbski, Anna Lewandowska, Magdalena Zioło.(2016).Fuzzy multi-objective modeling of effectiveness and user experience in online advertising, Expert Systems with Applications, Volume 65, 15 December 2016, Pages 315-331

Kim, C., Park, S., Kwon, K., & Chang, W. (2012). An empirical test to measure the effectiveness of online advertising in online marketplaces using a hierarchical Bayes model. .Expert Systems with Applications (39), 117–128

Kim, ji-Hern., Hyun, yong j., (2010). " A model to investigate the influence of marketing – mix efforts and corporate image on brand equity in the IT software sector " . industrial . Marketing Management. No 40,2011, pp 424-438, Emerald Group Publishing Limited

Kim, S., & Jun, J. (2016). The impact of event advertising on attitudes and visit intentions. .Journal of Hospitality and Tourism Management, 29, 1-8

Kirby, J. and Marsden .(2006). Connected marketing : the viral, buzz and word of mouth .revolution, Oxford : Butterworth-Heinemann, 2006

Madera. Quetzali, Oscar Castillo ⁎, Mario García-Valdez, Alejandra Mancilla..(2017). A method based on Interactive Evolutionary Computation and fuzzy logic for increasing the effectiveness of advertising campaigns, Information Sciences 414 (2017) 175–186

Marion Garaus and Udo Wagner , Anna-Maria B¨ack.(2017). The Effect of Media Multitasking on Advertising Message Effectiveness, Psychology & Marketing, Vol. 34(2):)138–156 (February 2017

Martín-Santana J.D., E. Reinares-Lara , P. Reinares-Lara .(2017).Influence of radio spokesperson gender and vocal pitch on advertising effectiveness: The role of listener gender, Spanish Journal of Marketing – ESIC, Volume 21, Issue 1, February 2017, Pages 63-71

McKay-Nesbitt, J., V. Manchanda, R., C. Smith, M., & A. Huhmann, B. (2011). Effects of age, need for cognition, and affective intensity on advertising effectiveness. Journal of Business Research (64), 12–17

Moch, Nils (2013). Competition in fragmented markets: New evidence from the German banking industry in the light of the subprime crisis. Journal of Banking &; Finance, Volume 37, Issue 8, Pp 2908–2919

Moorthy, S. and Hawkins, S. (2005). Advertising repetition and quality perception. Journal of .Business Research, 58, 354– 360

Mucan. Burcu, Özeltürkay. Eda Yaşa.(2014).Social Media Creates Competitive Advantages: How Turkish Banks Use This Power? A Content Analysis of Turkish Banks through Their Webpages, Procedia - Social and Behavioral Sciences, Volume 148, 25 August 2014, Pages 137-145

Muda, M.; Musa, R.; Niana-Mohamed, R. and Borhan, H. (2014). Celebrity Entrepreneur .Endorsement and Advertising Effectiveness. Social and Behavioral Sciences, 130, 11-20

Oladele. Kehinde Joseph, Ogunnaike, Akinbode Mosunmola, Aka Deborah .(2016). Effective Advertising: Tool For Achieving Client-Customer Relationships, Researchjournali's Journal of Media Studies Vol. 2 | No. 1 February | 2016

Olsen, G., Pracejus, J. W., & O'Guinn, T. (2012). Print advertising: White space. Journal of .Business Research (65), 855–860

Park, M. (2006). The compensatory effects of pictorial and verbal information for haptic information on consumer responses in non-store shopping environments. Unpublished Ph.D, .Ohio, Ohio state University

Park, T. (2008). Effective design of advertisement in handheld devices (Doctoral dissertation .)PURDUE UNIVERSITY

Park, T. (2008). Effectiveness Design of Advertisement in Handheld Devices. Dissertation for .doctoral degree, Purdue University, West Lafyette India

Pergelova, A., D. Prior and J. Rialp, 2010 Assessing Advertising Efficiency Does the Internet Play a Role? Journal of Advertising, 39(3): 39-54

Pilelienė Lina Viktorija Grigaliūnaitė , (2017)," The effect of female celebrity spokesperson in FMCG advertising: neuromarketing approach ", Journal of Consumer Marketing, Vol. 34 - .Iss 3 pp

Rosemary J. Averya, Matthew D. Eisenbergb, Kosali I. Simon .(2012).The impact of direct-to-consumer television and magazine advertising on antidepressant use, Journal of Health Economics, Volume 31, Issue 5, September 2012, Pages 705–718

Selviaridis. Kostas, Martin Spring b, Luis Araujo b .(2013). Provider involvement in business service definition: A typology, Industrial Marketing Management 42 (2013) 1398–1410

Strebinger .Andreas, Xiaoling Guo, Ferdinand Klauser, Peter Grant-Hayd .(2018).Is Multi-Ethnic Advertising a globally viable strategy for a Western luxury car brand? A mixed-method cross-cultural study, Journal of Business Research 82 (2018) 409–416

Suvittawat, A., (2015). "Commercial Bank Service Quality: Chonburi Province, Thailand". .International Journal of Management Sciences and Business Research. 4(2): 2226-8235

Tafani. Eric, Elyette Rouxa, Rainer Greifeneder .(2018).In the mood for action: When negative program-induced mood improves the behavioral effectiveness of TV commercials, Journal of Business Research 84 (2018) 125–140

Te'eni-Harari, T. (2014). "Clarifying the relationship between involvement variables and .advertising effectiveness among young people". Journal of Consum Policy. 37: 183–203

Thomas, S. and Kohli, S.C. (2011). Can Brand Image Move Upwards after Sideways? A .Strategic Approach to Brand Placements. Business Horizons, 54 (1): 41-49

Yahyaoui . Hamdi, Own .Hala S..(2018). Unsupervised clustering of service performance behaviors, Information Sciences 422 (2018) 558–571

Yuksel, A., Yuksel, F. & Bilim, Y. (2010) "Destination Attachment: Effects on Customer Satisfaction and Cognitive, Affective and Cognitive Loyalty", Journal Of Tourism Management, Vol.31, pp: 274-284.

STUDY OF NATURAL ANTIOXIDANT INHIBITORS FOR CORROSION PROTECTION OF MILD STEEL IN ACIDIC MEDIUM

Ms. Ibtihal Hassan AL-Bahrani, National University of Science and Technology, Sultanate of Oman.

ABSTRACT

Corrosion is a natural phenomena and it is a very big loss for structure as well as business; so corrosion minimization is one of the biggest areas of research. In this project; corrosion inhibition efficiency of berries juice extract was tested in mild steel sample against the sulfuric acid medium which was prepared from four types of berries; blackberries, blueberries, cranberries, and raspberries.

The experiment was designed by using Design Expert V 11 where immersed time, acid and berries juice extract concentrations were set as factors and a limited range was selected. The corrosion rate was set as a response. 20 runs were under taken and the experiments were done in the lab with maximum care.

The corrosion rate of 1020.374 mpy minimized to 597.659 mpy in presence of 5ml of extract and the efficiency of berries juice extract reached to 41.427%. The RSM gave 232.022 mpy corrosion rate as a minimum value and the optimum equation given by RSM was used for Genetic Algorithm analysis which gave 188.866 mpy corrosion rate at the optimizing conditions.

The surface and chemical characterization of the mild steel sample in presence and absence of berries juice extract were analyzed along with the relations between temperature, immersed time, acid and berries juice extract concentrations with response. The mixing tank was designed, calculations and drawing were presented.

KEYWORDS

Corrosion, Mild steel, Berries, Inhibitor, Weight loss, Sulfuric acid, RSM, SEM.

INTRODUCTION

Mild steel is used in most of our daily life applications as well as in industrial applications. In other hands, mild steel has low corrosion resistance which corrodes easily once it reacts with the surrounding environment; corrosion is a phenomenon where the mild steel reacts with the deteriorates and surrounding environment. Corrosion of mild steel is a continuous and constant problem where it severe financial and life loses as it leads to loss in the product, efficiency of equipment and failures in some cases (Roberge,2012).

The corrosion phenomenon controlled by adding a small amount of component to the corrosive environment which is a corrosion inhibitor. There are organic and inorganic inhibitors. The inorganic inhibitors are used in limit amount as they are toxic such as nitrites, molybate, nitrate, silicate, chromate, and phosphate (Cafferty,2010). Now a day, the corrosion engineers looking for cheap, renewable, friendly, readily available and ecologically acceptable inhibitors which are organic inhibitors which comes from the plants such as antioxidant components consisting of tannins, saponins, flavonoids, alkaloids and amino acids. Organic inhibitors consist of some heteroatoms such as oxygen, nitrogen, multi-bound, and sulphur atoms with the conjugated system which allows the absorption in the metal surface. Those atoms attract the absorption of some components on the metal surface and that done by the interaction of metals and by the electron of the conjugated system or by forming a bond with the free electron of heteroatom which forms a layer in the metal surface, as a result, it decreases the corrosion rate (Roberge,2012).

Berries are small pulpy fruits which have up to 9.2 mmol of antioxidant per 100g of berries. Berries can be used as a corrosion inhibitor as they are rich in antioxidant. The studies show that the berries juice extract of blueberries, cranberries, raspberries, and blackberries rich with antioxidant where the antioxidant capacity of them reach to 10 mg/ 100 g of blueberries (Michalska&Lysiak,2015), 58 mg/100g of cranberries (Blumberg,J.,et al,2013), 26.2mg/ 100 g of raspberries (Bobinaite,R.,Viskelis&Venskutonis,2015), and 21 mg/100 g of blackberries (Kaume,L.,et al,2012).

OBJECTIVES

To prepare anticorrosive extract (berries), to study corrosion in sample in presences and absences of the inhibitors, to determine the optimum conditions to inhibit corrosion and to design mixing tank.

METHADOLOGY, DESIGN AND EXPERIMENTATION

PREPARATION OF MILD STEEL SAMPLES

The mild steel sheets were cut into 2.5*3 cm^2 by using steel cutter and cleaned by surface cleaning machine to remove any corroded products in the metal surface then 4mm hole done by using drilling machine. Bunches for all metal samples (1-20) were made to differentiate between them. The drilled samples were washed by distilled water then rinsed with benzene and acetone respectively and dried with a hot air dryer. They were weighted using sensitive balance and kept in the desiccator until the time of usage.

PREPARATION OF CORROSIVE MEDIUM

The corrosive medium of different concentrations of H_2SO_4 prepared by using 36.8 N of the sulfuric acid.

PREPARATION OF BERRIES JUICE EXTRACT

137g, 209g, 282g, and 200g of raspberries, cranberries, blueberries, and blackberries respectively have been used to produce 309.5 ml of berries juice extract. The different types of berries washed by cold water, crashed in the electrical crasher and compressed mechanically by hand then filtered by cloth filter. The filtrate filtered again by using filter paper. The berries juice extract kept in the freezer at 0°C till the time of usage.

The efficiency of berries juice extract has been tested by running three experiments. The cleaned and weighted mild steel samples immersed in 0ml(blank), 2ml and 5ml of berries juice extract respectively with 50 ml of 1M H_2SO_4. After 2 hours; the samples have been taken out and cleaned by distilled water, benzene and acetone then dried by air dryer and weighted again. The corrosion rate and inhibitor concentration have been calculated by using equations (1) and (2) respectively.

$$Cr(mpy) = 22300\times\frac{\Delta W}{\rho\times A\times t} \quad \textbf{equation(1)}$$

$$IE(\%) = \frac{\text{initalcorrosionrate-finalcorrosionrate}}{\text{initalcorrosionrate}}\times 100 \quad \textbf{equation(2)}$$

DESIGN THE EXPERIMENT CONDITIONS

The experiments conditions have been designed by using Design Expert software-version 11 and following Response Surface Module(RSM); the range of the parameters have been given as inputs to the software which are inhibitor concentration (5-15 ml), immersed time (180-240 min) and acid concentration (1-2 M) and the response set as corrosion rate in mpy. Table(1) shows the experiment conditions given by the software.

Table(1): Designed runs given by RSM.

Run.No	Factor(A): inhibitor, ml	Factor(B): Time, min	Factor(C): Acid concentration, M
1	10	180	1
2	15	240	2
3	5	120	1
4	5	240	1
5	10	180	1.5
6	15	180	2
7	10	120	1.5
8	5	120	1.5
9	10	180	2
10	10	180	1.5
11	10	180	1.5
12	15	240	1
13	15	120	1
14	5	240	2
15	10	180	1.5
16	15	180	1.5
17	5	120	2
18	10	180	1.5
19	10	180	1.5
20	10	240	1.5

APPLYING THE DESIGNED RUNS IN THE LABORATORY

The weighted mild steel sample has been hanged by using a plastic thread in a 50 ml of sulfuric acid in presence of berries juice extract of known concentration and contact time as per the software; the batch has been kept in the shaker. After the contact time the mild steel sample cleaned by using 120 grade emery paper to remove any corroded products, washed by distilled water, rinsed by benzene and acetone respectively and dried by hot air dryer, weighted by using sensitive balance. The corrosion rate in mpy was calculated by using equation(1) where the weight losses results have been used.

STUDYING THE IMPACT OF TEMPERATURE IN CORROSION RATE

To study the impact of temperature in the corrosion rate; mild steel samples weighted and hanged for 2 hours in 50 ml of 1M sulfuric acid with 10ml of berries extract at different temperatures(30, 40, 50 and 60 °C) as shown in Table(2); each batch has been kept in the water bath shaker with heater to provide perfect contact between the solution and mild steel sample and to maintain the required temperature. Corrosion rate has been calculated by equation(1).

Table(2): Experiment conditions of temperature into corrosion rate.

SI.NO	Temperature (°C)	Inhibitor concentration (ml)	Acid concentration (M)	immersed time, (hours)
1	30	10	1	2
2	40	10	1	2
3	50	10	1	2
4	60	10	1	2

SURFACE AND CHEMICAL CHARACTERIZATION ANALYSIS

The mild steel surface of blank, absence and presence of berries juice extract samples have been studied by using Scanning Electron Microscopy (SEM) and Energy Dispersive Spectroscopy (EDS).

OPTIMIZATION STUDY BY GENETIC ALGORITHM

Matlab has been used for optimization where the genetic algorithm method has been followed. The actual equation form Design expert(RSM) has been used where "Optimtool" followed. The same limit of the conditions from RSM has been selected. The same previous procedure has been followed in the lab with maximum care by using those optimum conditions which given by the software, 1.001M of H_2SO_4 with 15 ml of berries juice extract immersed for 237.208 minutes.

DESIGN OF MIXING TANK

Design of mixing tank is a part of the project where the industrial scale mixing tank has been designed, the tank will be formed from high resistance material such as S.S to be able to resist the corrosive medium of sulfuric acid which will be mixed with berries juice extract. It is a container which used to mix various material (heterogeneous materials) together to form homogenous materials even the different viscosity materials can be mixed easily which provides continuous process, good control, easy cleaning, and simplest construction. It can be made from S.S material which provides a smooth surface and it can be easily cleaned (Petropedia,2018).

Parameters to be considered to design mixing tank are length, torque, weight, stiffness and hydraulic loading.(Dynamixianc,2015). The important factors in design mixing tank are tank type, tank volume, specific gravity, process and environment.(Ibryamoval,Y.et al,2009)

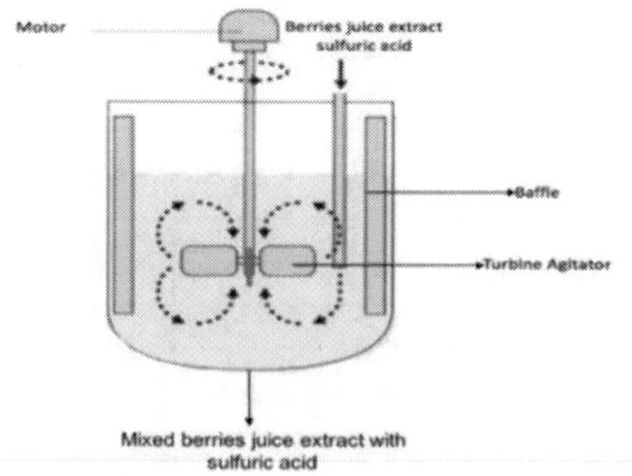

Figure(1): Mixing Tank(*Mixing Tank*,2018)

From the graph in Figure(2) Turbine agitator has been selected as it is suitable with the viscosity of the fluid and the tank volume. (Sinnot&Towler,2009)

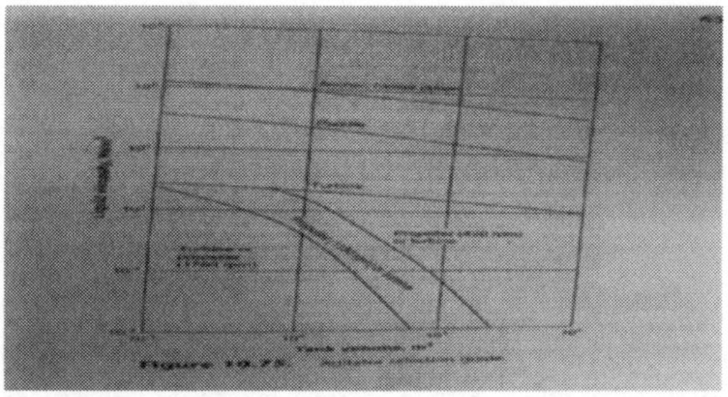

Figure(2): Graph of liquid viscosity Vs tank volume. (Sinnot&Towler,2009)

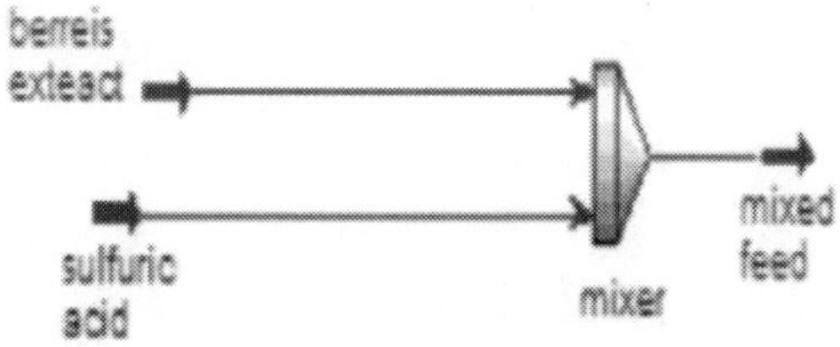

Figure(3): Result of Mixing Tank by Aspen Hysys.

Table(3): Calculated parameters for designing mixing tank

Tank diameter	m	3.5
Tank Height	m	1.81
Across sectional area	m^2	9.62
Volume of the tank	m^3	17.5
Impeller diameter	m	1.16
Length of paddle	m	0.29
Width of paddle	m	0.23
power	watt	204085
Reynolds Number	-	136486.6391

RESULTS AND ANALYSIS

EFFECT OF BERRIES JUICE EXTRACT CONCENTRATION INTO CORROSION RATE AND INHIBITOR EFFICIENCY

As shown in Table(4) the corrosion rate in the absence of berries juice extract inhibitor reach to the maximum (1020.374mpy); the corrosion rate has been minimized to 898.117mpy and 597.659mpy in presence of 2 and 5 ml of berries juice extract inhibitor respectively which means that berries juice extract inhibits the corrosion as the concentration of the berries juice extract increases. As well as the inhibitor efficiency reaches to 41.427% which obtained at 5ml of berries juice extract inhibitor in 1M of sulfuric acid immersed for 2 hours. The adsorption of the organic compounds such as ascorbic acid that present in the berries juice extract is blocking the active sites on the mild steel samples as it is forming a protective film on the metal surface as what kumar.A,2012 mention in his studies so the corrosion rate keeps decreasing as the amount of the berries juice extract is increasing. (Kumar.A,2012) Those compounds have heteroatoms such as oxygen which is a center of adsorption. In addition; It forms a coordinate bond as it has the ability to transfer the electrons to the vacant d-orbital of mild steel from the heteroatoms. (Ostovari, L. et al,2009)

Table(4): Experimental results of testing the efficiency of berries juice extract.

SI.NO	Inhibitor concentration (ml)	Corrosion rate (mpy)	Inhibitor efficiency (%)
1	0	1020.374	-
2	2	898.117	11.976
3	5	597.659	41.427

EFFECT OF IMMERSED TIME INTO CORROSION RATE AND INHIBITOR EFFICIENCY

The maximum corrosion rate reaches to 765.005mpy which obtained at 5ml of berries juice extract in the presence of 50 ml of 2 M of H_2SO_4 at 2 hours. Since the minimum corrosion rate of 232.022mpy obtained in presence of 5ml of berries juice extract with 50 ml of 1 M of H_2SO_4 immersed for 4 hours. The effect of immersed time into corrosion rate can be observed at low acid concentration and in the presence of berries juice extract, the corrosion rate is minimized at high immersed time. The decrease in the corrosion rate with the increase in immersed time is related to the adsorption of the corrosion product ($FeSO_4$) as shown in equation(3); According to Xianghong,L. et al ; the corrosion product of $FeSO_4$ becomes thicker with the immersed time, so it form a protective film (barrier) in the surface of mild steel (Xianghong, L. et al,2009).

$$Fe_{(s)}+H_2SO_{4(aq)} \rightarrow FeSO_{4(aq)}+H_{2(g)}$$, **equation(3)**

EFFECT OF ACID CONCENTRATION INTO CORROSION RATE

The effect of acid concentration into corrosion rate has been studied by using 1, 1.5 and 2 M of H_2SO_4 at different inhibitor concentrations and immersed time where the corrosion rate increase with the acid concentration at low inhibitor concentration and immersed time which reach to 765.005mpy at 2M H_2SO_4, 5ml of berries juice extract which immersed for 120 minutes as a maximum corrosion rate as shown in Table (3). In presence of the corrosive medium of H_2SO_4 with less amount of berries juice extract; the mild steel surface was exposed to H_2SO_4 medium where the corrosion reaction started as shown in equation(3) and $FeSO_4$ was formed with H_2 bubbles.

EFFECT OF TEMPERATURE INTO CORROSION RATE

From Table(5) can be noted that the rate of corrosion increases with temperature where it reaches to maximum 2407.994mpy at maximum temperature of 60°C. According to Ambrish, S. et al, at high temperature the berries juice extract molecules were partially absorbed in the mild steel surface so it was exposed to the corrosive medium of H_2SO_4, as a result, increasing the corrosion rate (Ambrish,S. et al,2012).

Table(5): Results of studying the effect of temperature into corrosion rate

SI.NO	Temperature,° C	Corrosion rate, mpy
1	30	701.507
2	40	1666.404
3	50	2008.861
4	60	2407.994

ANALYSIS BY RSM

Once the experimental results and corrosion rate calculated the RSM analysis has been done where the corrosion rate has been given to the software as a response. The response must be modified where the results fit with linear. The fit summary shows that the model F-value and P-value of 12.66 and 0.0001 respectively imply that the model is significant where A, B, C, and B^2 are the significant terms in this model only. 1.48 is a lack of fit P-value which indicates that the fit is not significant as there are pure errors.

0.5562 of the predicted R^2 is a reasonable agreement with 0.7105 of the adjusted R^2. The signal to noise ratio of 13.7252 of Adeq precision that the ratio grater than 4 is desirable. (Mycourses,2016)

The process variables are studied using ANOVA. The estimate coefficients in term of coded factors are A-Inhibitor, B-time, C-Acid and B^2. It is creating an accompanying condition for the sample with 1.0000 correlation coefficient. The coded and actual equations given by software are shown in equation(4) and (5) respectively.

- Corrosion rate(mpy)= 370.52-40.52(A)-126.97(B)+ 80.85(C) +66.58$(B)^2$ **equation(4)**
- Corrosion rate(mpy)= 1189.15487-8.10479(Inhibitor) -8.77421(Time) +161.69302 (Acid Conc)+0.018495 $(Time)^2$ **equation(5)**

The various relations can be studied from one factor graph; it can be noted that the corrosion rate increases with acid concentration and decreases with the increase in the amount of berries juice extract and immersed time as can be seen from the plots in Figure(3):Plot pf corrsoion rate Vs inhibitor concntration, Figure(4): plot of corrosion rate Vs acid concentration and Figure(5):Plot of corrosion rate Vs immersed time respectively.

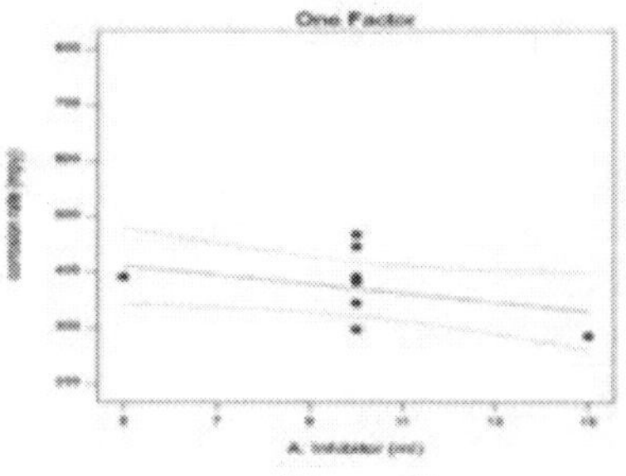

Figure(3)

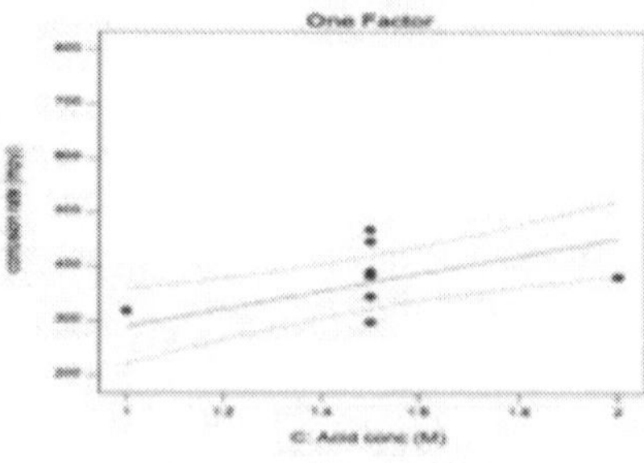

Figure(4)

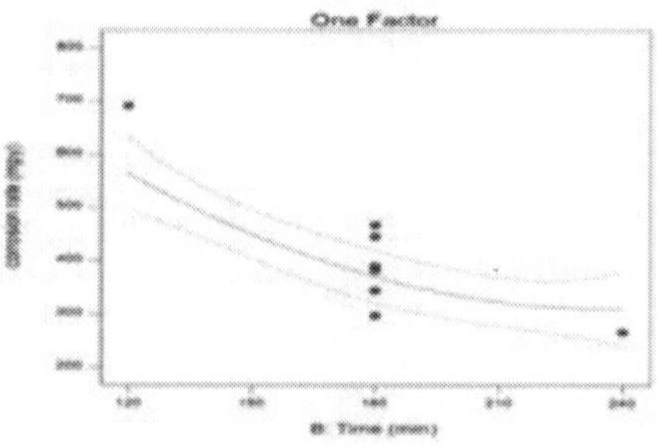

Figure(5)

The effects of two factors into the corrosion rate can be studied from the 3D plots as shown in Figure(6): 3D plot of corrosion rate Vs inhibitor concntration and acid concntration and Figure(7): 3D plot of corrosion rate Vs inhbitor concntration and immersed time.

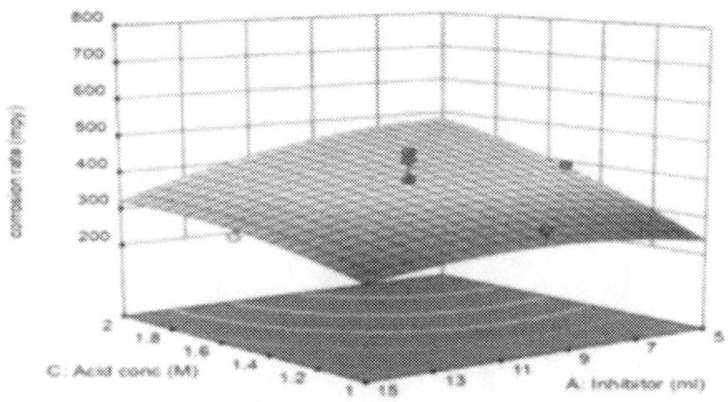

Figure(6)

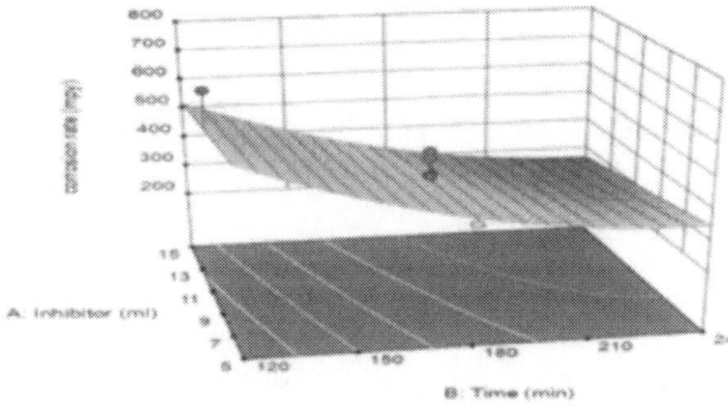

Figure(7)

The optimum conditions given by the RSM of 1M of H_2SO_4 with 5ml of berries juice extract inhibitor runs for 240 minutes gives 213.402mpy as a minimum corrosion rate but the experimental results at the same conditions gives 232.022mpy which means that they are too close as the suggested solution considered as an excellent optimum conditions.

SEM ANALYSIS

The SEM has been done for the blank mild steel sample for facilitating comparison as shown in Figure(8) which shows smooth mild steel surface compare to others. Rough surface in absence of berries juice extract is shown in Figure(9); where corrosion starting with the formation of H_2 bubbles as per the equation(3) which means that the surface of mild steel is exposed to the H_2SO_4 medium. In presence of 15 ml of berries juice extract, the surface of mild steel is protected by the adsorption of the organic compounds that are present in the berries juice extract where fewer cracks and pits can be noted and smooth surface is shown in Figure (10), which means that the mild steel surface is protected. (Raja, P. et al, 2008)

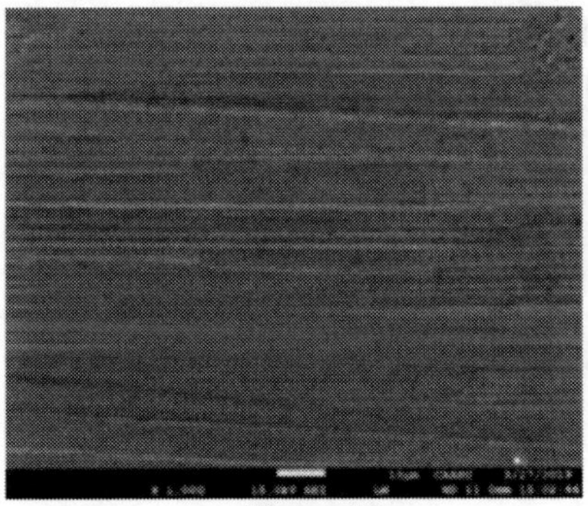

Figure(8)

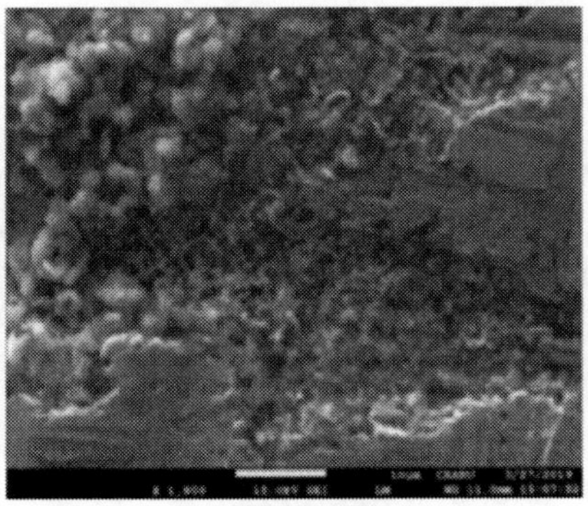

Figure(9)

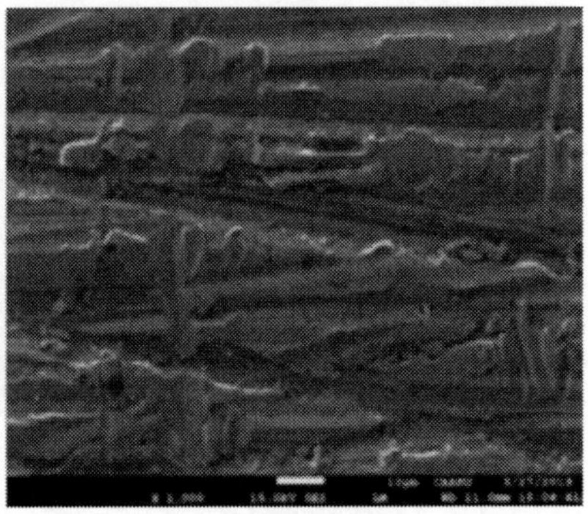

Figure(10)

EDS ANALYSIS

EDS has been used to study the chemical characterization or elemental analysis; Figure (11) shows the blank mild steel sample before being immersed in 1M H_2SO_4 in absence of berries juice extract where high Fe peak is shown, which indicates the presence of high amount of iron which reaches to 88 Wt% in the mild steel surface. Figure (12) shows the mild steel surface that has been immersed for 2 hours in 50 ml of 1M H_2SO_4 in absence of berries juice extract; the mild steel surface contains low amount of Fe (87.3 Wt%) compared to the blank mild steel sample. In Figure (13) the mild steel sample immersed for 2 hours in 50 ml of 1M H_2SO_4 in the presence of 5ml of berries juice extract, it shows high spark of Fe where the mild steel surface contains lower amount of Fe (84.8 Wt%) compared to Figure(11) and (12) which indicate lower iron dissolution rate in 1M H_2SO_4 medium containing berries juice extract (Alaneme, K. et al,2016). A high amount of oxygen and carbon are observed in presence of berries juice extract compared to the absence of berries juice extract. Si is shown in the absence and presence of berries juice extract which means that's it comes out to the surface during the reaction with sulfuric acid.

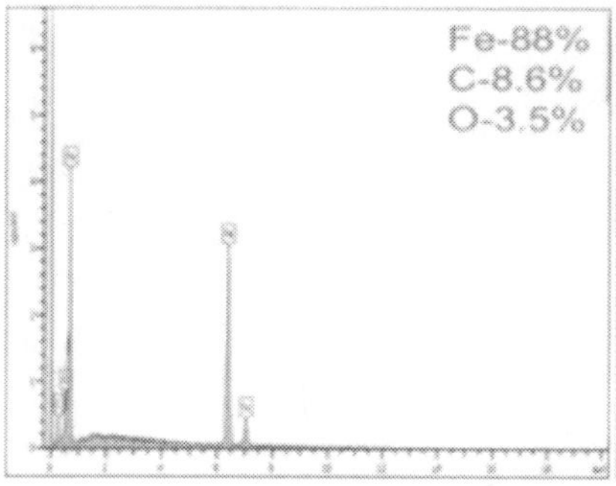

Figure(11)

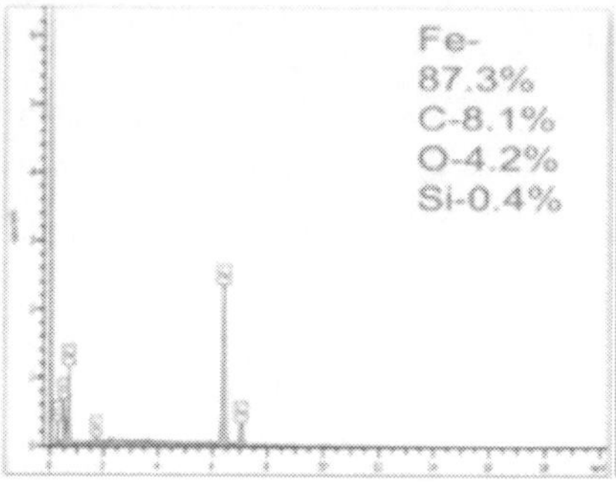

Figure(12)

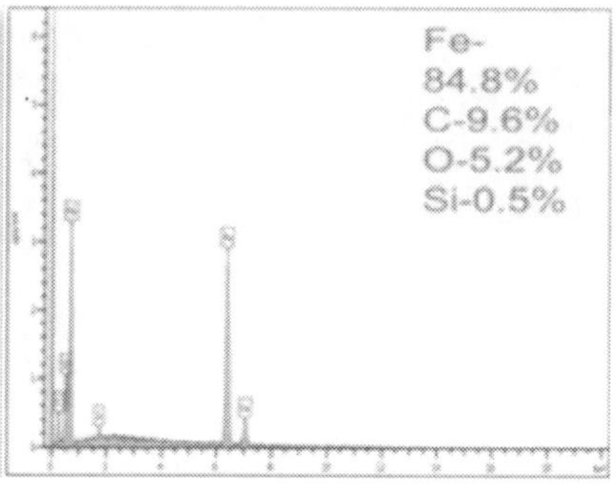

Figure(13)

GENETIC ALGORITHM OPTIMIZATION

The experimental result gives 288.367mpy as a corrosion rate. The software gives the minimum corrosion rate value of 188.866mpy from 50 runs as shown in Figure(14). A small difference can be noted as the genetic algorithm is not the exact fit with linear equations so the experimental and software results are too close to each other.

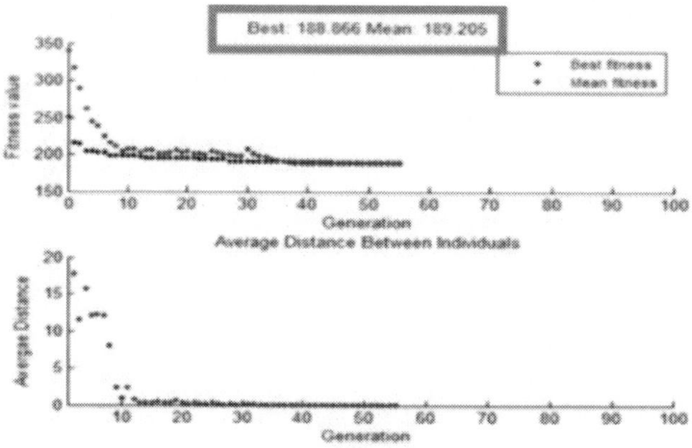

Figure(14): Genetic Algorithm optimization.

CONCLUSION

In this project, all the objectives have been achieved by designing and experimenting.
Firstly; the anticorrosive extract (berries) has been prepared, the prepared extract which contains 4 types of berries namely blueberries, blackberries, cranberries, and raspberries inhibit the mild steel corrosion in sulfuric acid medium with 41.427% efficiency in presence of 5 ml of berries extract.
Secondly, the mild steel corrosion in the presence and absence of berries juice extract were studied where the corrosion rate reaches to 1020.374mpy in absence of berries juice. By using 5 ml of berries juice extract the corrosion rate minimized to 232.022mpy.
Thirdly, the optimum conditions to inhibit the mild steel corrosion were studied by Design Expert V11. At the optimum conditions of 1M of H_2SO_4. 5ml of berries juice extract, 240 minutes and at room temperature the corrosion rate reaches to 232.022mpy as a minimum value.
The actual equation obtained from Design Expert V11 was later optimized using Genetic Algorithm. Genetic Algorithm shows that at 1.001M of H_2SO_4, 15 ml of berries juice extract and 237.208 minutes at room temperature, the corrosion rate is minimized to 188.866mpy.
Finally, mixing tank has been designed with 1.18m height and 2.5m diameter and a turbine agitator has been selected.

ACKNOWLEDGEMENT

I would like to thank god for giving me patience and power to work in my most important part of my life, my father, mother, my brother and my friends for their help and support. I would also like to express my sincere gratitude to Dr. Priy Brat Dwivedi for his support and help in developing my project idea, Mr. M.J.Varghese for his help in RSM analysis, Mr. Hatem Yazidi and Dr. Moammar Ali AlTubi for providing me the technical knowledge in Matlab optimization, Mr.Mohammed Salim Adeeb in providing me the technical help in the laboratory, Ganesh Nagraj Patil, Dr.Lakkim Setty

Rao, and Dr.Santosh Madhav Walke for their help in the calculations of design mixing tank. Last but not the least, I would like to thank National University of Science and Technology-College of Engineering for giving me this opportunity to complete the project in their laboratories and supporting me by providing me with the required chemicals.

REFERENCES

1.Alaneme,K. et al .,2016.Corrosion Inhibitory Properties of Elephant Grass(Pennisetum Purpureum) Extract Effect on Mild Steel Corrosion in 1M HCl Solution :*Alexandria Engineering Journal*.[e-journal]. 55 (8). p. 1069-1076. Available from: https://www.sciencedirect.com/science/article/pii/S1110016816300242.[Accessed: 8th May 2019].

2.Ambrish,S. et al .,2012. The Effect of Environmentally Bening Fruit Extract of Shahjan(Moringa Olifera) on the Corrosion of Mild Steel in Hydrochloric Acid Solution:*Journal of Chemical Engineering Communications*.[e-journal]. 199 (6). p. 63-77. Available from:https://www.tandfonline.com/doi/abs/10.1080/00986445.2011.570390.[Accessed: 6th May 2019].

3.Blumberg,J.,et al.,2013.Carnberries and Their Bioactive Constituents in Human Health:*Journal of Advances in Nutrition*.[e-journal]. 12 (7). p. 618-632. Available from: https://academic.oup.com/advances/article/4/6/618/4595566.[Accessed: 1th May 2019].

4.Bobinaite,R.,Viskelis,P.&Venskutonis,P.,2015.Chemical Composition of Raspberry Cultivars:*International Journal of Molecular Sciences*.[e-journal]. 16 (8). p. 18642-18663. Available from: https://www.researchgate.net/publication/306158828_Chemical_Composition_of_Raspberry_Rubus_spp_Cultivars.[Accessed: 1th May 2019].

5.Cafferty,E.,2010.*Introduction to Corrosion Science*.United States:Springer.
Roberge,P.,2012.*Handbook of Corrosion Engineering*.2nd edition.United States:McGraw-Hill company.
6.Dynamixianc,2015.*Optimal Mixer Design* .[Online]. Available from: https://www.dynamixinc.com/optimal-mixer-design. [Accessed: 14th December 2018]

7.Ibryamoval,Y. et al .,2009.Design of Mixers for Chemical and Biochemical Industry: *Trakia Journal of Sciences*.[e-journal]. 8 (6). p. 6-11. Available from: http://tru.uni-sz.bg/tsj/Vol8.Suppl.3.2010/Y.Ibryamova.pdf.[Accessed: 13th December 2018].

8.Kaume, L. et al.,2012. The Blackberry Fruit A Review on Its Composition and Chemistry, Metabolism and Bioavailability and Health Benefits:*Journal of Agricultural and Food Chemistry*.[e-journal]. 37 (7). p. 5716-5727. Available from: https://www.ncbi.nlm.nih.gov/pubmed/22082199.[Accessed: 6th May 2019].

9.Kumar.A,2012. Inhibition of Mild Steel Corrosion in HCl Solution Using Pipali(Piper Longum)Fruit Extract:*Arabian Journal for Science and Engineering*.[e-journal]. 38 (9). p. 85-97. Available from: https://link.springer.com/article/10.1007/s13369-012-0409-9.[Accessed: 9th May 2019].

10. Michalska,A.& Lysiak.G .,2015.Bioactive Compounds of Bluberries:*International Journal of Molecular Sciences*.[e-journal]. 16 (8). p. 18642-18663. Available from: https://www.ncbi.nlm.nih.gov/pmc/articles/PMC4581264/.[Accessed: 1th May 2019].

11.*Mixing Tank*,2018.[Online]. Available from: https://www.conceptdraw.com/examples/agitator-tank-symbol.[Accessed: 19th November 2018].

12. Mycourses,2016. *Response Surface Modelling-RSM*.[Online]. Available from:https://mycourses.aalto.fi/pluginfile.php/364340/mod_resource/content/1/BOS_Lecture3_2016.[Accessed: 4th May 2019].

13. Ostovari, L. et al.,2009.Corrosion Inhibition of Mild Steel in 1M HCl Soluition by henna extract:*Journal of Corrosion Science*.[e-journal]. 32 (9). p. 1935-1949. Available from: https://www.sciencedirect.com/science/article/pii/S0010938X09002157. [Accessed: 6th May 2019].

14. Petropedia,2018. *Mixing Tank*.[Online]. Available from: https://www.petropedia.com/definition/7742/mixing-tank.[Accessed: 11th December 2018].

15.Raja,P. et al.,2008.Inhibitive Effect of Black Pepper Extract on the Sulphuric Acid Corrosion of Mild Steel:*Journal of Materials Letters*.[e-journal]. 62 (9). p. 1977-2979. Available from: https://www.sciencedirect.com/science/article/abs/pii/S0167577X08001316.[Accessed: 8th May 2019].

16. Sinnot,R.&Towler,G.,2009.*Chemical Engineering Design*.5th edition.India:Elsevier.

17. Xianghong,L. et al.,2009.Inhibition Effect of 6-benzylaminopurine on the Corrosion of Cold Rolled Ssteel in H_2SO_4 Soluition:*Journal of Corrosion Science*.[e-journal]. 51 (9). p. 620-634. Available from: https://www.sciencedirect.com/science/article/pii/S0010938X09000080. [Accessed: 6th May 2019].

590434 Verification of Firefighters' Heuristics and Predictive Model Development through Big Data Analysis

Abstract:

The heuristics accumulated in the field activities of firefighters are reviewed through big data analysis of fire occurrences and emergency responses in Gyeonggi-do Province, Korea and expanded as predictive models to be utilized for proper fire prevention activities according to time, day, and target through quantitative modeling. Empirical rules with high sympathy are collected through direct interviews with firefighters. Among these, three types of high-priority rules are compared and analyzed for big data such as fire occurrence and damage history in Gyeonggi-do in 2018. Big data comparison analyses are conducted, including the number of fires and damages that occurred in Gyeonggi-do in 2018. Furthermore, fire occurrence patterns by region, day of the week, time of day, and building type are derived. In addition to using empirical rules that have been validated through research, relatively inexperienced firefighters also can make better decisions by relying on refined quantitative predictive models replacing empirical rules, including local government and time-based factors that reflect big fire occurrence data.

CONVERTING SHELTER-IN-PLACE LOCATIONS TO ZERO VULNERABILITY SAFE HAVENS

J. A. Rau
MineARC Systems America, LLC
4850 W. Ledbetter Drive
Dallas, Texas 75236
USA

A. Young
MineARC Systems, Pty Ltd
274 Welshpool Rd
Welshpool, WA 6106
Australia

M. McDermott
High Performance Building Solutions
7415 Folger Dr.
Charlotte, NC 28226
USA

CONVERTING SHELTER-IN-PLACE LOCATIONS TO ZERO VULNERABILITY SAFE HAVENS

ABSTRACT

While it is not uncommon to find shelter-in-place or safe haven locations designated across chemical facilities, what varies is the level of protection offered at these sites. Functionality can range from merely a muster point (shelter-in-place) to a positively pressurized room (safe haven). Either way, when determining the room's acceptable performance criteria, the industry generally applies ERPG-3[1] or a vulnerability factor. An acceptable vulnerability factor is often deemed the probability of a fatality being less than one percent.

In 2017, MineARC Systems and High Performance Building Solutions (HPBS) began working with a leading energy company to convert five shelter-in-place spaces to safe havens, which included additional life support systems that would ensure zero percent vulnerability. The objective of this paper is to examine the process to harden the space against contaminated air intrusion and what critical life support systems are essential for converting a shelter-in-place location to a safe haven.

INTRODUCTION

Emergency shelter-in-place (SIP) forms an integral part of any plant's broader emergency response plan. Explosions, fires, and the release of smoke or other toxic gases are some of the types of incidents that can occur despite high levels of planning and safety precautions[2].

Numerous government agencies and other organizations recommend shelter-in-place as a measure to reduce harm in the event of a chemical release. Any building may be used as a temporary measure to reduce health risk from exposure to the toxic materials by simply closing windows and doors and turning off ventilation fans and air conditioning/heating systems.

Three factors govern the effectiveness of a SIP location: 1) the tightness of the building or leakage rate, 2) the concentration of the toxic gas, and 3) the release duration. The tightness of the building is primarily based on building construction; however, temperature differential and wind speed also play a factor. Identification of the maximum toxic concentration occurs during a Facility Siting Study (FSS). FSSs are a hazard analysis that defines Maximum Credible Event (MCE) scenarios using a consequence-based approach to produce a quantitative risk assessment. Determining potential consequences is a necessary step in the process of developing a comprehensive safety plan. The FSS determines the expected release duration based on stored inventories of toxic materials. The higher the toxic concentration and longer the release duration, the less effective sheltering in place becomes.

While occupants may feel safe sheltering-in-place from the external toxic hazard, typical buildings have high leakage rates. A typical wood framed constructed building can have 3-5 air changes per hour (ACH50[3]), rendering it ineffective against most chemical releases. Additionally, sealing to an acceptable airtightness can be time consuming and expensive and even with professional air sealing it is very difficult to air seal existing structures to an acceptable tightness.

A case study for the conversion of five shelter-in-place spaces to zero percent vulnerability safe havens is presented. There are considerations for effectively sealing existing infrastructure and ensuring positive pressure to ensure contaminants do not infiltrate safe haven locations along with the critical element of life support. The critical elements of life support that are of primary importance to human survival in a safe haven are:

- Positive pressure
- Breathable air supply
- Supplemental O_2
- Removal of accumulated CO_2

---

[1] Emergency Response Planning Guidelines (ERPGs) are values developed by the AIHA Guideline Foundation's Emergency Response Planning Committee. ERPG-3 is the maximum airborne concentration below which it is believed nearly all individuals could be exposed for up to 1 hour without experiencing or developing life-threatening health effects.

2 The CSB is an independent federal agency charged with investigating industrial chemical accidents. Over the last 10 years they have completed investigations into 15 separate chemical release incidents.

[3] ACH50 is the number of times the air volume in a building changes per hour at 50 Pascal of pressure. Commercial air leakage rates are commonly reported at 75 Pascals, but this leakage study was conducted at 50 Pascals based on the targets included in the FSS report.

- Cooling and dehumidifying

CASE STUDY

In 2017 MineARC Systems and its project partner HPBS began working alongside a major energy company to convert five shelter-in-place locations into zero percent vulnerability safe havens.

The process started with the energy company completing its FSS, which included plume modeling to define the chemical release concentrations at each building location along with ideal shelter-in-place locations[4] and internal minimum positive pressure[5]. This information was used by MineARC to calculate indoor concentrations and occupant vulnerability to highlight that shelter-in-place alone would result in the vulnerability of greater than zero percent.

From there, HPBS conducted air infiltration testing following ASTM E779-10 to determine an air leakage baseline (see Table 6) for rooms that were designated to be sealed (pre-weatherization). A site assessment was then completed to review each location and determine its suitability for sealing and conversion to a safe haven. Firstly, the measured baseline infiltration rates were compared to the FEMA 453 estimates based on construction type[6] (see Figure 2). All surfaces of the room were examined for ease of conventional air sealing and mechanical equipment sealing. Rooms that included false floors, large amounts of electrical conduit, cabinetry, computer servers or windows etc. were deemed difficult to seal (i.e. control rooms). Secondly, the building or rooms open floor space was checked to ensure it could accommodate the specified number of people in an emergency. FEMA 453 section 1.6 recommends 20ft^2 per person for durations of less than 24 hours. With the FSS recommending 4 hours, it was deemed acceptable to have an unrestricted floor space of 6ft^2 per person.

Manual weatherization (air sealing) of the rooms included installing a cleanroom suspended ceiling system with gaskets and tiles that met the air permeance rate for an air barrier material, cleanroom airtight light fixtures with gaskets, sealing around electrical and plumbing, air sealing all visible air pathways to make sure the interior gypsum board was functioning as the air barrier, installing low-leakage HVAC dampers, installing an emergency shut off for the mechanical system and adding door weather-stripping and sweeps. Post-weatherization air leakage tests were conducted to determine the air infiltration rate of the SIP spaces after completion of the manual air sealing, cleanroom suspended ceiling system and HVAC work. The weatherization air sealing methods improved the air barrier integrity for all the proposed SIP spaces by between 57% and 65% (refer to Table 6).

Aerosol-based envelope sealing was used to reduce leakage rates even further. AeroBarrier® is a building envelope sealing technology that simultaneously measures and seals buildings. AeroBarrier® works by forcing an atomized sealant into orifices under pressure gradually sealing leaks. Rooms were sealed during night shift with personnel able to re-access the rooms by morning. The AeroBarrier® air sealing methods improved the air leakage for all the SIP spaces between 7% and 88% over the post manual weatherization leakage rates (refer to Table 7). AeroBarrier® is not effective for sealing large holes so manual air sealing is required for certain locations prior to using AeroBarrier®.

MineARC engineers once again calculated infiltration concentration rates along with the vulnerability rate. As the probability of a fatality was still higher than zero, MineARC then specified its AirBANK Pressure System and necessary carbon dioxide and oxygen control equipment.

SAFE HAVEN LIFE SUPPORT EQUIPMENT

Primary Life Supporting Atmosphere

A safe haven needs to provide a reliable breathable atmosphere independent of external toxic gas concentrations. Human breathing consumes oxygen (O_2) and expels carbon dioxide (CO_2), carbon monoxide (CO), water vapor, and heat. At complete rest, an average person exhales air at 17% O_2 and 3.2% CO_2. This breathing rate results in a 'resting' O_2 uptake of 0.3L/min of O_2 and a resting CO_2 discharge of 0.24L/min (Brake, Fellows, & Bates, 1999). Taking into consideration the above values, a person in a closed system such as a shelter-in-place succumbs to CO_2 poisoning well before O_2

4 Shelter-in-place location is beyond the scope of this paper, but should be determined based on the chemical hazard, facility layout, and location of personnel.

5 Positive pressure requirements are based on average wind speed at the facility. As wind speed increases so does the SIP leakage rate.

6 The FEMA 453 estimates are intended to estimate filter sizing when ASTM E779 air leakage testing is not conducted.

asphyxiation. Venter et al. (1998) found an average oxygen consumption for 12 entrapped individuals of 0.44 liters per minute, at STP, over 24 hours (including sleeping). A figure of 0.5 liters per minute over a non-sleeping entrapment period of 12 hours is a prudent, conservative design value. This corresponds to a metabolic rate of about 160W or 80W/m^2 for a typical person with a body skin area of about 2m^2, which is equivalent to a breathing rate of about 12.5 liters per minute of fresh air.

Consequently, life support systems are needed within a safe haven that can ensure positive pressure keeps contaminants outside, along with replacing the O_2 consumed, and removing exhaled CO_2, moisture, and heat to maintain the atmosphere within acceptable threshold limits.

Positive Pressure

Positive pressure is integral to prevent the ingress of airborne contaminants that may compromise the breathable atmosphere for occupants. Establishing a Class 1 Safe Room (FEMA 453) requires the sufficient release of filtered air from an external source, to form a positively pressurized zone.

Buildings are typically constructed to provide natural or forced ventilation to allow permanent occupancy and potentially combined with large exposed windows and surfaces which can present challenges for ensuring protection during a chemical release. When determining adequate positive pressure within a space, several structural factors need to be considered, including:

- Shelter volume
- HVAC (Heating, Ventilating, and Air Conditioning)
- Door and window tightness
- Existing opening such as vents and elevator shafts
- Indoor and outdoor temperature
- Wind speed and building orientation
- The external screening of the structure

Building design and multiple ingress points suggest the conversion of a smaller interior room within a building can provide a higher level of protection, than the whole building, as it is less subject to wind and buoyancy forces which induce the infiltration of external hazards.

Breathable air from compressed air cylinders venting directly into the SIP forms a positive internal pressure. Determining the amount of air required to reach the necessary level of positive pressure involves measuring the areas internal volume and the relationship between the internal and external atmospheric conditions.

After reaching required ACH (air changes per hour) or tightness, a positive-pressure management system must be sized for the expected shelter duration.

Breathable Air Supply

If the shelter-in-place is airtight, a supply of breathable air is necessary to support occupants over the required shelter duration. Unventilated SIPs that are tightly sealed cannot be occupied for long periods without the risk of high carbon dioxide levels. A secondary source of breathable air must be circulated within the shelter to meet occupancy over the required amount of time.

Clean, breathable air is supplied to a chemical, biological and radiological (CBR) safe room in a few ways: by trapping air inside the room and minimizing the air exchange, passing contaminated air through a filter (not practical for high concentration chemical releases or for chemicals that cannot be filtered) to purify it as it is supplied to the room, and the supplemental supply of oxygen through high pressure compressed air cylinders.

The air supply must contain enough oxygen in its mix, and not contain contaminants like carbon monoxide, carbon dioxide, water, or any other toxins that poison safe haven occupants during entrapment.

Table 1 lists international standards for flowrates for breathable compressed air for self-contained breathing apparatus (SCBA), supplied air respirators (SAR) and refuge chambers.

Table 1 – Standards for acceptable air supply requirements

Air supply requirements	Relevant Standard
Open-circuit positive-pressure SCBA, including all components and accessories except the air cylinder (shell) must resist the permeation and penetration of HD and GB chemical agents when tested for a period of 6 hours on an upper-torso manikin connected to a breathing machine operating at an airflow rate of 40 liters per minute (lpm), 36 respirations per minute, 1.1 liters tidal volume.	NIOSH 42 CFR Part 84 Respiratory Protective Devices
The flow rate should be at least 85L/min (3cfm) per person to remove expired CO2 and CO but at least 255L/min (9cfm) to provide adequate cooling and maintain comfortable conditions. The air inflow is assumed to be at a temperature of 26.5°C and 65% relative humidity	ITA Report No. 14 – Guideline for the provision of refuge chambers in tunnels under construction, May 2014
For manually operated blower type air-hose respirators, the blower shall be capable of delivering air to the space enclosed by the facepiece at a flow rate of at least 85 L/min.	AS/NZS 17:16:2003 Respiratory Protective Devices

The NIOSH standard of 40L/min is for SCBA apparatus (i.e. mouthpiece or face mask) where dilution of CO_2 is not a concern. Conversely, the ITA and Australian standards specify 85L/min and are concerned with both oxygen consumption and carbon dioxide dilution. It is important to note that three times the amount of air is required to maintain comfortable conditions inside a refuge chamber, when compared to simply breathing (Health & Safety Executive UK, 2006).

Unless the positive pressure system is supplying breathable air at a rate that maintains oxygen and compressed gas levels within acceptable ranges (dependent on safe haven occupancy and duration), a CO_2 scrubber and supplementary oxygen supply are required.

Supplemental O_2

O_2 is an essential requirement for cellular activity, and therefore, the absence of O_2 for relatively short periods results in death. Humans within an enclosed environment consume O_2 and if not replaced, would result in an irrespirable atmosphere. An O_2 concentration of less than 11% poses a high risk of death.

Retaining the oxygen levels between acceptable limits is critical. Various international standards outline acceptable O_2 limits for a breathable environment; examples are shown in Table 2.

Table 2 - Standards for allowable O_2 levels

Minimum O_2 Content (% by volume)	Maximum O_2 Content (% by volume)	Relevant Standard
18.5%	23.0%	Mines Safety and Health Administration 30 CFR Parts 7 and 75 Refuge Alternatives for Underground Coal Mines; Final Rule
19.5%	23.5%	Occupational Safety and Health Administration 29 CFR 1910.146 Permit-Required Confined Spaces for General Industry
19.5%	23.5%	Australian Standard AS 2865–2009 Safe working in a confined space

OSHA and Australian Standards for working in a confined space specify an acceptable lower limit of 19.5%, which is consistent with many other standards around the world. The slightly lower MSHA limit of 18.5% is based on a person at rest, not working, therefore, not likely to have trouble in breathing; conditions that are likely to be present in a refuge chamber.

O_2 itself is not flammable but supports and accelerates combustion. OSHA defines O_2 enriched mixtures or atmospheres as those containing more than 23.5% O_2 by volume (OSHA, 2013). In O_2-enriched atmospheres, the reactivity of O_2 significantly increases the risk of ignition and fire.

Within a safe haven, pure O_2 must be utilized in conjunction with a CO_2 removal system.

Removal of Accumulated CO_2

Protection within a safe haven is substantially increased by filtering CO_2 from the air. CO_2 is being produced continuously by the cells of our body and is a byproduct of respiration. The inability to expel this gas or an environment with high concentrations of CO_2 leads to CO_2 poisoning. Breathing air with more than 3% CO_2 can induce dizziness, headaches, and vision disturbance. Prolonged exposure at greater than 6% could result in unconsciousness and death. Listed in Table 3 are examples of international standards of satisfactory CO_2 removal requirements for a breathable environment.

Table 3 – Acceptable CO_2 removal requirements

CO_2 Removal Requirements	**Relevant Standard**
CO_2 removal components shall remove CO_2 at a rate of 1.08 ft^3 (30.5 L)/h/person.	Mines Safety and Health Administration 30 CFR Parts 7 and 75 Refuge Alternatives for Underground Coal Mines; Final Rule
OSHA has established a Permissible Exposure Limit (PEL) for CO2 of 5,000 parts per million (ppm) (0.5% CO2 in air) averaged over an 8-hour workday (time-weighted average or TWA.)	Occupational Safety and Health Administration
5.2.8 Atmospheric monitoring. Atmospheric monitoring There should be the capability to monitor temperature continuously, and for carbon monoxide, carbon dioxide, and oxygen in the chamber during use in both externally supported or standalone modes. The concentration of carbon dioxide should not exceed 10000 ppm.	ITA Report No. 14 – Guideline for the provision of refuge chambers in tunnels under construction, May 2014

The CO_2 that is exhaled by the occupants within the safe haven must be removed by, constant flushing by an external breathable air source (e.g. compressed air cylinders used for positive pressure) or by the regeneration of the internal atmosphere with a CO_2 absorbent (scrubber).

Different regulations around the world, state varied human production rates between 30–50L per hour per person. When considering an acceptable production rate; it must be noted that the rate of breathing is primarily triggered by the CO_2 content of inspired air with higher CO_2 levels triggering faster respiration rates (Brake, Fellows, & Bates, 1999) (i.e. if the maximum allowable CO_2 concentration inside the safe haven is 1%, a lower production rate is acceptable compared to a 3% CO_2 limit).

If CO_2 removal mechanisms are inadequate, the occupants can develop CO_2 toxicity by re-breathing their own exhaled CO_2.

Cooling and Dehumidifying

Thermal tolerability with a safe haven should be addressed during the design phase. Assumptions must be made in preparation for safe haven design in regard to the number of personnel, activity level, available hydration and heat production. The minimum airflow to maintain a comfortable environment is set by taking these assumptions into account as well as the input temperature and humidity of supplied air.

The healthy human body maintains its internal temperature around 37°C. An essential requirement for continued normal body function is that the deep body temperature is maintained within the narrow 1°C limit (Epstein, Moran, 2006). A change of body temperature exceeding 1°C occurs only during illness or when environmental conditions surpass the body's ability to cope with extreme temperatures.

When a person is subject to extreme heat and humidity over some time, the person may experience a very rapid progression of symptoms or exhibit a few of the less significant symptoms before falling victim to more severe forms of heat-related illness.

The acceptable cooling requirements for underground refuge chambers are defined in various international standards, examples of which are listed in the below table.

Table 4 – Standards for acceptable cooling requirements

Cooling Requirements	Relevant Standard
Simulated emergencies, in which a full complement of people has occupied a refuge chamber for a significant period, indicate that humidity and temperature can increase very rapidly to potentially heatstroke-inducing levels. Refrigerated air conditioning is strongly recommended for both externally supported and stand-alone refuge chambers to counter this potentially dangerous problem. Inevitably, this places a heavy demand on the stand-alone power supply, but there are systems available that can cope, and at an acceptable cost.	Department of Mines, Industry Regulations and Safety, 2005 Refuge Chambers in Underground metalliferous mines – guideline: Safety and Health Division, Department of Industry and Resources Western Australia, 32 pp
Apparent temperature in the fully occupied refuge chamber must not exceed 95 °F.	Mines Safety and Health Administration 30 CFR Parts 7 and 75 Refuge Alternatives for Underground Coal Mines; Final Rule
It should be possible to maintain an internal temperature not exceeding 30 °C wet bulb globe temperature for at least 24 hours	ITA Report No. 14 – Guideline for the provision of refuge chambers in tunnels under construction, May 2014

MSHA regulations specify a maximum internal apparent temperature[7] of 95°F (35°C). NIOSH specifies an equivalent maximum wet bulb temperature of 28.9°C (NIOSH, 2007).

The temperature maintained inside the safe haven is determined by the metabolic heat liberated by the occupants, dissipated to the external environment by the heat transfer characteristics of the room or building. The external temperature primarily influences the heat transfer while the humidity is the combination of moisture produced by the occupants (via exhaled air and sweat) and water liberated by CO_2 scrubbing chemicals.

It is important to remember that in most cases, before sheltering-in-place, the air conditioning system will need to be shut down (unless it is a split system). For short durations, artificial cooling is not a critical requirement unless external temperatures are extraordinarily high, and/or the safe haven's volume is small compared to its occupancy.

INFILTRATION TESTING

Background

High Performance Building Solutions Inc. (HPBS) was contracted to perform an air infiltration study for spaces identified as candidates to be used as shelter-in-place (SIP) in the event of an accidental chemical release. This SIP strategy uses the existing structure and its indoor atmosphere to separate individuals from a hazardous outdoor atmosphere temporarily.

A series of pre-weatherization air leakage tests were conducted to determine the baseline air infiltration rate of potential SIP spaces. Manual air sealing, a cleanroom suspended ceiling system and low-leakage HVAC dampers were installed to reduce the overall air infiltration of the SIP spaces and to close off air leakage pathways from non-protected spaces in the building through the HVAC system. Post-weatherization air leakage tests were conducted to determine the air infiltration rate of the SIP space after the completion of weatherization and HVAC work. The SIP spaces were additionally air sealed using the proprietary AeroBarrier® system to further reduce infiltration rates.

Scope

Air sealing work was completed, and low-leakage HVAC dampers were installed to reduce the overall air infiltration of the SIP spaces and to isolate the HVAC from non-protected spaces in the building. Post-weatherization air leakage tests were conducted to determine the air infiltration rate of the SIP space after the completion of air sealing, suspended ceiling and HVAC work. AeroBarrier® air sealing was conducted on SIP spaces as a pilot project on September 18 through September 21, 2017. Post AeroBarrier® air leakage tests were conducted on September 21 and September 22, 2017, to determine the air infiltration rate of the SIP space after the completion of the AeroBarrier® sealing.

7 Apparent temperature is the temperature equivalent perceived by humans, caused by the combined effects of air temperature and relative humidity.

Following ASTM E 779-10, pressurization air leakage tests were conducted on the following spaces recommended in the Facility Siting Study (FSS).

Table 5: SIP specifications

SIP Space	Occupancy	Floor Area (Square Feet)	Volume (Cubic Feet)
SIP Space 1	40	990	10,481
SIP Space 2	6	955	8,210
SIP Space 3	5	391	3,917
SIP Space 4	10	1,697	16,968
SIP Space 5	40	293	2,637

Air Leakage Test Results

Positive pressure air leakage test results are reported in four different ways:

CFM4 is the quantitative volume of air passing through an enclosure under a pressure difference of 4 Pascals (equivalent to a 5mph wind). This is a measurable physical quantity and the value recommended in the FSS to eliminate toxic gas ingress into the SIP spaces.

CFM25 is the quantitative volume of air passing through an enclosure under a pressure difference of 25 Pascals (0.1-in w.g.). FEMA 453 requires Class 1 Safe Havens be pressurized to 25 Pascals. This is a measurable physical property.

CFM50 is the quantitative volume of air passing through an enclosure under a pressure difference of 50 Pascals (0.2-in w.g.).

ACH50 is the quantitative volume of air passing through an enclosure under a pressure difference of 50 Pascals (0.2-in w.g.) normalized by the enclosure volume. ACH is the abbreviation for air changes per hour. It is the number of times the air volume in a building changes per hour at 50 Pascals of pressure.

Table 6 below shows the CFM50 and ACH50 for each of the SIP spaces pre-weatherization and post-weatherization along with the percentage reduction in leakage. SIP Space 1 and SIP Space 5 did not have pre-weatherization results as initially the whole building was tested. Smaller spaces were selected as the whole building was too large to feasibly convert to a SIP space (re-testing of the new SIP spaces was not completed). The control room, SIP space 4, was abandoned in favor of a smaller adjacent room due to high leakage post manual weatherization and the computers that could not be shut down, which prevented the use of AeroBarrier® in the space.

Table 6: Post-weatherization leakage results

Safe Haven Space	CFM50		ACH50		% Reduction
	Pre-Weatherization	Post-Weatherization	Pre-Weatherization	Post-Weatherization	
SIP Space 1	N/A	1,871	N/A	10.7	N/A
SIP Space 2	3,623	1,560	26	11.4	57%
SIP Space 3	385	135	6	2.1	65%
SIP Space 4	N/A	N/A	N/A	N/A	N/A
SIP Space 5	N/A	655	N/A	19.4	N/A

The AeroBarrier® air sealing was prescribed a target. According to the AeroBarrier® simultaneous air leakage calculations, all the spaces met and surpassed this airtightness target; AeroBarrier® sealing was halted at this point. Table 7 shows the CFM50 and ACH50 for each of the SIP spaces as tested by HPBS post AeroBarrier® along with the percentage reduction. The HPBS results were utilized to calculate the amount of positive pressure required to convert the rooms to positive pressure safe havens.

Table 7: Post AeroBarrier® leakage results

Safe Haven Space	CFM50		ACH50		% Reduction
	Post-Weatherization	Post AeroBarrier®	Post-Weatherization	Post AeroBarrier®	
SIP Space 1	1,871	455	10.7	2.6	76%
SIP Space 2	1,560	182	11.4	1.3	88%
SIP Space 3	135	125	2.1	1.9	7%
SIP Space 4	N/A	173	N/A	3.3	N/A
SIP Space 5	655	138	19.4	3.1	79%

The AeroBarrier® air sealing methods improved the air barrier integrity for all the proposed SIP spaces between 7% and 88% over the post manual weatherization leakage rates. Air leakage was still observed behind many electrical penetrations including HVAC thermostats, and fire and smoke alarms. These were covered with plastic to protect them from the AeroBarrier® Sealant but had hidden holes behind them.

Figure 1 displays the post AeroBarrier® leakage rate in comparison to the FEMA 453 estimates for air leakage rates based on construction types. FEMA 453 estimates are reported in square feet of floor area at 25 Pascals and do not take into consideration the ceiling height and volume of the space.

Figure 1: Post AeroBarrier® results compared to FEMA 453 estimates

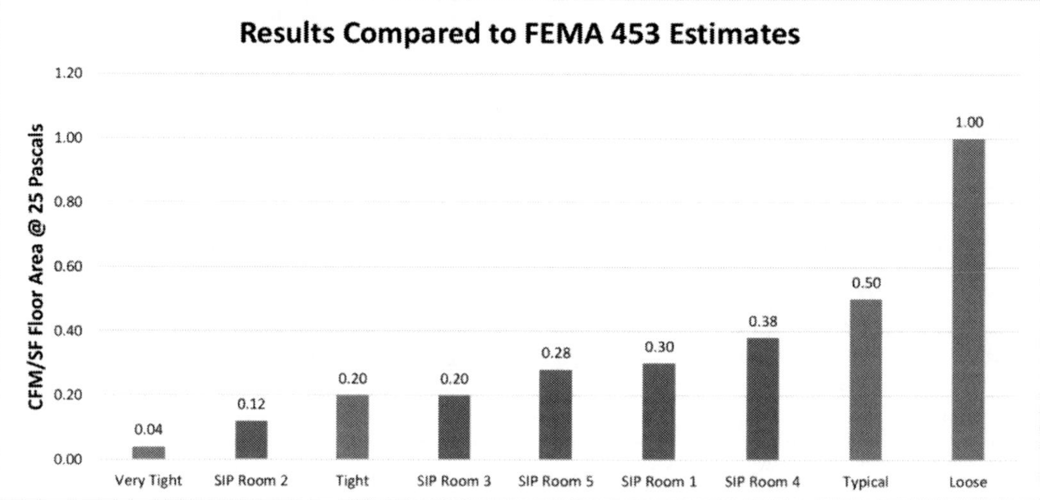

Figure 2 is an estimated leakage rate based on construction type to estimate the filtration and positive pressure system size under the FEMA 453 standard. All the SIP locations were between tight and typical except for SIP Room 2, which was a small prefabricated metal constructed portable building.

Figure 2: Estimated leakage rate per square foot of floor space to achieve 0.1 iwg (25Pa) overpressure

Construction Type	cfm per square foot of floor area
Very tight: 26-inch thick concrete walls and roof with no windows	0.04
Tight: 12-inch thick concrete or block walls and roof with tight windows and multiple, sealed penetrations	0.20
Typical: 12-inch thick concrete or block walls with gypsum wall board ceilings or composition roof and multiple, sealed penetrations	0.50
Loose: Wood-frame construction without special sealing measures	1.00

Table 8 shows the combined AeroBarrier®, and manual air sealing methods improved the airtightness for all the proposed SIP spaces between 68% and 95% over the original leakage rates.

Table 8: Air infiltration improvement for SIP spaces

Safe Haven Space	ACH50			% Reduction
	Pre-Weatherization	Post-Weatherization	Post AeroBarrier®	
SIP Space 1	N/A	10.7	2.6	76%
SIP Space 2	26.5	11.4	1.3	95%
SIP Space 3	5.9	2.1	1.9	68%
SIP Space 4	N/A	N/A	3.3	N/A
SIP Space 5	N/A	19.4	3.1	84%

Figure 3 compares the ACH50 pre-weatherization, post-weatherization and AeroBarrier® air infiltration results.

Figure 3: Pre-weatherization, post-weatherization and AeroBarrier® comparison

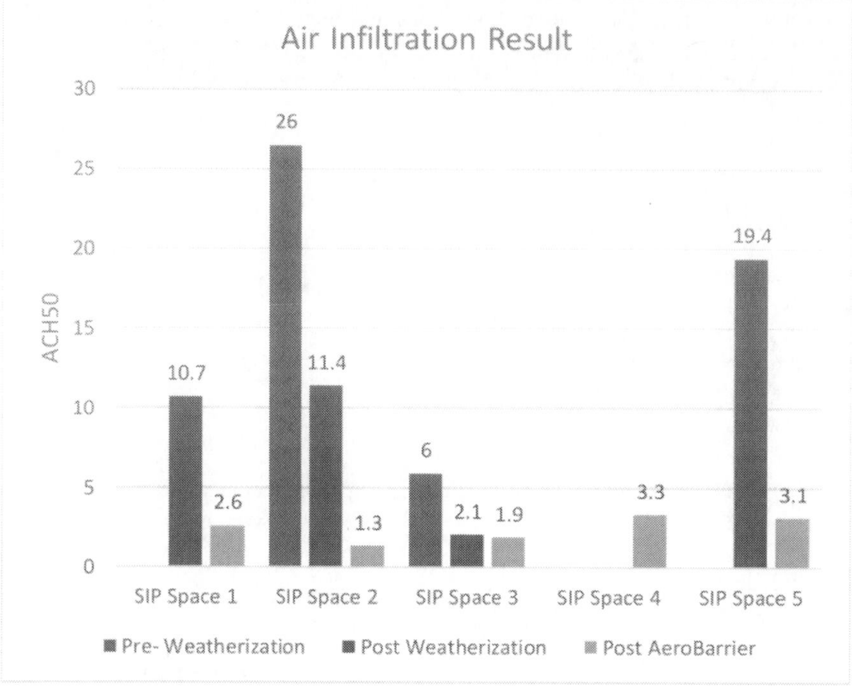

Life Support Systems

MineARC re-calculated the vulnerability for each SIP space based on the post AeroBarrier® leakage rates. All SIP spaces still had a vulnerability of 100% and therefore required the installation of independent life support systems to be deemed zero percent vulnerability safe havens.

Figure 4 shows the toxic gas concentration, duration of the release, and vulnerability for each SIP space. As the largest MCE was a full tank rupture the toxic concentrations were extremely high, but the duration was short at one minute.

Figure 4: SIP space vulnerability post AeroBarrier®

	TOXIC GAS				
Safe Haven Location	Worst Case Threat Scenario	Toxic External [ppm]	Duration of Reslease [min]	ACM	Vulnerability
SIP Space 1	Tank Rupture	585,000	1	0.032	100%
SIP Space 2	Tank Rupture	652,000	1	0.042	100%
SIP Space 3	Tank Rupture	660,000	1	0.007	100%
SIP Space 4	Tank Rupture	412,000	1	0.042	100%
SIP Space 5	Tank Rupture	600,000	1	0.028	100%

The leakage rate at 4 Pascals of pressure difference is slightly more than would occur under a 5 mile per hour wind used in the dispersion model in the FSS report based on climatic data for the facility (to maintain positive pressure and neutralize air infiltration). Post-weatherization infiltration rates in cubic feet per minute at 4 Pascals of pressure for each of the SIP spaces is shown in Table 9.

Table 9: Post AeroBarrier® air leakage results at CFM4

Safe Haven Space	Volume (Cubic Feet)	CFM4
SIP Space 1	10,481	99
SIP Space 2	8,210	32
SIP Space 3	3,917	24
SIP Space 4	3,110	45
SIP Space 5	2,637	21

The air leakage test results provide a mathematical formula that can be used to calculate the amount of breathable compressed air cylinders (437ft^3) that must be added to each candidate SIP space to maintain a 4 Pascal positive pressure to overcome a 5mph design wind coincident with a toxic release.

MineARC Systems was contracted to install life support systems to convert the SIP spaces to positive pressure safe havens. The MineARC AirBANK Positive Pressure System provides rapid pressurization, which is activated and maintained using the AirBANK Control via a simple human-machine interface (HMI) touch screen. MineARC's integrated Aura-FX Gas Monitor ensures breathable air automatically remains within acceptable limits. Additionally, the installation of supplementary oxygen and carbon dioxide scrubbing systems in safe havens, are as needed.

Figure 5 shows the number of 437 cubic feet cylinders required for each safe haven location. The minimum breathable air per person is 2.5cfm (70L/min) per person based on MineARC's experience and testing within the refuge chamber industry.

Figure 5: Quantity of cylinders required for 4Pa positive pressure

	SAFE HAVEN SPECIFICATIONS		POSITIVE PRESSURE (@ 4Pa)						
Safe Haven Location	Occupancy	Duration	CFM	CFM per Floor Space	ACH	No. Cylinders	No. Cylinder Racks	Breathable Air per Person [CFM]	Breathable Air per Person [LPM]
Safe Haven Space 1	40	4	99	0.10	0.6	55	5	2.5	70.8
Safe Haven Space 2	6	4	32	0.03	0.2	18	2	5.4	152.9
Safe Haven Space 3	5	4	24	0.06	0.4	14	2	4.8	135.9
Safe Haven Space 4	10	4	45	0.14	0.9	25	3	4.5	127.4
Safe Haven Space 5	40	4	21	0.07	0.5	12	2	0.6	17.0

The maximum CO_2 concentration is at 1% and O_2 between 19.5% and 23.5% based on OSHA limits (refer to Table 2). Due to the volume and number of people in Safe Haven Space 5, it required supplemental oxygen and carbon dioxide scrubbing to maintain the internal gas concentrations within specified limits. Three sodium chlorate oxygen candles and two AirGEN CO_2 scrubbers were included in the design.

Figure 6: Internal gas concentrations in safe havens

	SAFE HAVEN SPECIFICATIONS		POSITIVE PRESSURE (@ 4Pa)			02 & CO2 CONCENTRATIONS (@ 4Pa)			
Safe Haven Location	Occupancy	Duration	CFM	CFM per Floor Space	ACH	Indoor O_2 %	No. O_2 Candles	Indoor CO_2 %	No. AirGEN Scrubbers
Safe Haven Space 1	40	4	99	0.10	0.6	20.45	0	0.50	0
Safe Haven Space 2	6	4	32	0.03	0.2	20.84	0	0.17	0
Safe Haven Space 3	5	4	24	0.06	0.4	20.77	0	0.23	0
Safe Haven Space 4	10	4	45	0.14	0.9	20.66	0	0.31	0
Safe Haven Space 5	40	4	21	0.07	0.5	18.67	3	2.04	2

In 2018 MineARC completed installation and training on the five safe haven spaces. Each safe haven was tested during the commissioning process to ensure it could maintain acceptable positive pressure.

CONCLUSIONS

The conversion of shelter-in-place (SIP) locations to zero percent vulnerability safe havens is a four-step process consisting of the following:

1) Identifying the correct SIP location based on ease of sealing and small volume based on the required occupancy.
2) Manual air sealing of SIP spaces.
3) AeroBarrier® sealing of SIP spaces.
4) Installation of life support equipment.

Identifying SIP spaces that can seal from the outset is crucial to a successful result. In the case study, three of the five original locations were abandoned in favor of rooms in adjacent locations due to high leakage rates or reducing the size to make the project more cost effective. Control rooms pose the most difficulty in trying to seal due to the false floor and a myriad of cable tray and pass-throughs installed for the computer systems. The smaller the room and the more robust the construction (i.e. prefabricated metal or concrete buildings), the better it seals.

Manual air sealing needs to be overseen by someone with familiarity in building tightness. While most of the tasks are general construction, standard contractors are not trained to work at the level of precision needed to ensure proper sealing. Even a tiny orifice or ceiling tile not sealed correctly can ruin a leakage test result. HPBS and MineARC Systems have developed a system for ensuring airtightness in construction methodology and by avoiding un-sealable spaces or spaces that are not cost effective to seal.

For future construction projects, incorporation of SIP spaces should occur during the design phase. Air barrier commissioning is standard in new construction and should include the following tasks at a minimum: design review of the plans with specifications for the air barrier and mechanical system - specifically as these relate to spaces designated for use as safe havens; quality assurance inspections during construction to ensure a continuous air barrier is in place; and final air leakage testing of the continuous air barrier to ensure the leakage rate of the safe haven spaces are within acceptable limits.

Complementary life support equipment is essential to sustain life within an airtight environment. For personnel to remain within the sealed safe haven, the environment must be habitable. Air is to be free of toxins such as carbon dioxide and acceptable levels of oxygen maintained. Positive pressure systems reduce the ingress of toxins and provide some or all of the required source of breathable air. Additional equipment such as compressed air, oxygen cylinders and air scrubbers are used to meet the shortfall. Furthermore, internal conditions must consider the human condition, supporting workable body temperature and humidity.

Reducing the risk of harm to personnel is a priority for emergency response planning. While the likelihood of an MCE occurring may be low, the risk to on-site personnel in the event of an incident is high. Converting shelter-in-place (SIP) locations to zero vulnerability safe havens improves emergency response.

REFERENCES

ASTM E779-10, *Standard Test Method for Determining Air Leakage Rate by Fan Pressurization*, ASTM International, West Conshohocken, PA, 2019, www.astm.org

Brake, Fellow, & Bates (1999). *Criteria for the design of emergency refuge stations for an underground metal mine*. Aus. IMM Proceedings, No.2, 1999, pp 1–7.

Department of Mines and Petroleum, 2013, *Refuge chambers in underground mines — guideline: Resources Safety*, Department of Mines and Petroleum, Western Australia, 41 pp.

Epstein, Y., & Moran D. (2006). *Thermal Comfort and the Heat Stress Indices,* Industrial Health, 44, 388–398.

FEMA 453. (2006). *Design Guidance for Shelters and Safe Rooms.* Federal Emergency Management Agency

Health & Safety Executive UK (2008). *Guidance and Information on the role and design of safe havens in arrangements for escape from mines.* Retrieved from http://www.hse.gov.uk/pubns/mines08.pdf

ITA Working Group 5 (2016). *Provisions for refuge chambers in tunnel projects under construction.* ITA Report N°014 - V2 / March 2018

Mines Safety and Health Administration (2008). *30 CFR Parts 7 and 75 Refuge Alternatives for Underground Coal Mines; Final Rule.*

National Institute for Occupational Safety and Health (2007). *Research Report on Refuge Alternatives for Underground Coal Mines.* Retrieved from http://www.cdc.gov/niosh/docket/archive/pdfs/NIOSH-125/125 ResearchReportonRefugeAlternatives.pdf

Occupational Safety & Health Administration (2013). *Permit-required confined spaces.* Retrieved from http://www.osha.gov/pls/oshaweb/owadisp.show_document?p_table=STANDARDS&p_id=9797

Suming Du. (2014). *On the effectiveness of shelter-in-place as a measure to reduce harm from atmospheric releases. Journal of Emergency Management, Vol. 12, No, 3, May/June 2014.*

Venter. J. (1998). *Portable Refuge Chambers: Air or Tomb in Underground Escape Strategies.* School for Mechanical and Materials Engineering, Potchefstoom University, Republic of South Africa

Worksafe Australia National Standard (2005). *Australian Standard AS: NZS 1716:2003 Respiratory protective devices.*

Yantosik. G. (2006). *Shelter-in-Place Protective Action Guide*. The Chemical Stockpile Emergency Preparedness Program (CSEPP). U.S. Department of Energy

589008 How the Field of Human Factors Helps Solve the Riddle of Serious Injuries and Fatalities

Abstract:

Process and occupational safety are often treated as distinct endeavors, with the former tending to focus on facility system function and the latter on individual worker wellbeing. Though other differences certainly exist between the two, both endeavors are generally purposed to prevent incidents that can cost lives, health, time, and money, in addition to employee morale and public trust. The overarching thesis of this paper is that the field of Human Factors is uniquely positioned to effect positive change in both of these major safety categories.

The backdrop for our argument is the recent finding that while injury and illness incidence rates published by the Bureau of Labor statistics have decreased in recent years, the occurrences of serious injuries and fatalities have declined at a significantly slower rate. We discuss possible explanations for this finding but note that regardless of the preferred explanation, the trend should motivate scientists, engineers, and other safety professionals to revisit long-held accident causation theories holding that near misses – these are incidents that could have resulted in serious injuries and fatalities but did not – are precursors of future serious incidents. While this idea still has merit, the just-mentioned statistics warrant reconsideration of the relations between serious and less-serious incidents to explain why the rates of the two are not decreasing in lockstep. We suggest these data indicate additional (human) factors are involved.

To address this riddle and to support our general thesis, we (a) review core tenets of process safety, occupational safety, and the field of Human Factors, (b) point out that the human is always an element common to both process and occupational safety contexts, and (c) discuss how preventing or mitigating human error can benefit both. We offer specific examples from both safety contexts to show how Human Factors perspectives and techniques can help. Examples touch on standards, policies, and other administrative controls (e.g., procedures), human-machine interfaces of both workplace hardware (e.g., valves) and software (e.g., SCADA systems), environments (e.g., physical, regulatory, and cultural), and communications between personnel (e.g., elements of team resource management). We conclude that fundamental Human Factors considerations can help prevent and mitigate human performance gaps and therefore increase the rate of reduction in serious injuries and fatalities in both process and occupational safety.

589233 Initial Conditions Used in Atmospheric Dispersion Models:Issues with Commonly Used Assumptions for Gas Jets and Plumes

Abstract:

When simulating jets and plumes, all atmospheric dispersion models require initial conditions that appropriately parameterize the behavior of a gas release to the atmosphere. Often, process conditions are such that loss of primary containment of a gas will result in choked flow conditions. A great deal of experimental and theoretical consideration has been given to the flow from stagnation conditions to choked flow conditions, but less experimental work has centered on describing the flow through the depressurization process to atmospheric pressure where atmospheric dispersion models would be applicable. Traditional modeling of the depressurization process has taken the simplified approach of assuming a (depressurized) "top hat" profile (laterally constant) with application of the principles of conservation of mass, momentum, and energy. Such assumptions are known to result in predicted depressurized velocities that exceed the (sonic) velocity at the point of loss of containment, and because of the energy balance, the associated temperatures are significantly less than temperatures at the choked conditions. Physically, these effects could possibly self-cancel because as jet velocity slows, the decrease in kinetic energy of the jet will increase the jet temperature. However, no atmospheric dispersion model used for routine hazard assessment purposes accounts for kinetic energy in the energy balance. The traditional modeling approaches of the depressurization process are reconsidered based on analysis of available data, and an alternative approach to modeling the depressurization process is proposed. Example cases showing the important differences between the two approaches are discussed.

Fault Tree Uncertainty Analysis

Raymond "Randy" Freeman
S&PP Consulting
12303 Lake Shore Ridge
Houston, TX 77041
Cell: 713 408 0357
Rafree@yahoo.com

For Presentation at the 2020 Spring National AIChE Meeting
Houston, Texas

March 30, 2020

ABSTRACT

Fault tree analysis (FTA) is a widely used methodology in the process industries. FTA is used for the development of failure mechanisms, computation of failure frequencies and the determination of the probability of failure on demand of safety systems. Much of the data used in a FTA study are uncertain. For example, the failure rate of a pump is often not known with great precision. Likewise the failure rates of instrumentation are often known only within some defined limits. The common practice, used by analysts in the quantification of a fault tree, is to use the most likely or best guess as to the needed failure rate data. The use of best guess values as data inputs to the quantification of a fault tree creates uncertainty in the computed results.

This paper presents a general methodology for the determination of the impact of uncertainty on the results of a fault tree study. The general methodology is based on the mathematics of propagation of error and variance contribution analysis. For a fault tree that contains 10 event minimal cut sets, the methodology will compute the variance in the probability (or frequency) of the top event with a maximum error of 5%. For 5 event minimal cut sets the error in variance of the top event probability (or frequency) will be no more than 2%. An example is presented to illustrate the application of the fault tree uncertainty analysis methodology to a real world problem.

Keywords

Quantitative Risk Assessment, Fault Tree Analysis, Uncertainty Analysis

1. Background

Fault tree analysis is a widely used method for representation of failure mechanisms in chemical plants. The methodology is described in the CCPS CPQRA guideline book [Ref. 1] and is extensively discussed in Chapter 9 of Lees [Ref. 2]. Lacking in these discussions is a systematic method to quantify the uncertainty of the results of the fault tree calculations.

Fault tree analysis has also been used in the aircraft and aerospace industries, nuclear power plant review and is recognized by the OSHA Process Safety Management (PSM) standard [Ref. 3] as a method for process hazards analysis.

2. Review of Error Propagation and Variance Contribution Analysis (VCA) Methodology

The mean and variance of a function of random variables can be approximated using the method described by Haugen [Ref. 4] and applied by Freeman [Ref. 5, 6, 7]. Define an arbitrary function of a set of random variables, x_i, as:

Let

$$Y = F(x_i) \qquad \text{(Eq 1)}$$

The mean of Y can be estimated using the following approximation:

$$E(Y) = F[\, E(x_i)] \qquad \text{(Eq 2)}$$

Where:

E(Y) = expected value of random variable Y = mean of Y

$E(x_i)$ = expected value of random variable x_i = mean of x_i

The variance of Y can likewise be estimated as:

$$V(Y) = \sum_{i=1}^{n} \left[\frac{\partial Y}{\partial x_i} \right]^2 V(x_i) \qquad \text{(Eq 3)}$$

Where:

$V(Y)$ = variance of random variable Y as defined above in Equation 1

$V(x_i)$ = variance of random variable x_i as defined above in Equation 1

Note that the variance is simply the square of the standard deviation. Using the variance will simplify the mathematics that is described below. The contribution of each independent random variable to the overall variance in the function is:

$$V(Y \text{ from } x_i) = \left[\frac{\partial Y}{\partial x_i}\right]^2 V(x_i) \qquad \text{(Eq 4)}$$

The relative contribution of each term to the overall variance V(Y) is a measure of the importance in the uncertainty in the particular random variable, x_i. In effect, this is a sensitivity analysis combined with a uncertainty evaluation. The variance contribution combines the sensitivity in the answer to changes in the uncertain random variable, x_i, with a measure of the uncertainty in the random variable, x_i. The overall variance in Y is found by summing the sensitivity weighted variances from each random variable.

Numerical Estimate of the Sensitivity

Let us return to the fundamental definition of the derivative.

As before, set y = F(xi)

$$\frac{\partial y}{\partial x} = \lim_{\Delta xi \to 0}\left[\frac{F(xi+\Delta xi) - F(xi)}{\Delta xi}\right] \qquad \text{(Eq 5)}$$

We are interested in a numerical estimate of the derivative. Therefore, we will make the following approximation.

$$\frac{\partial y}{\partial x} \sim \left[\frac{F(xi+\Delta xi) - F(xi)}{\Delta xi}\right] \qquad \text{(Eq 6)}$$

Linear Functions

Let us examine the approximation for a simple linear function.

Let Y = AX + B (Eq 7)

What is the sensitivity of Y to a change in variable X? By simple calculus:

$$\frac{\partial y}{\partial x} = A \qquad \text{(Eq 8)}$$

We can also find the sensitivity using the numerical approximation as:

$$\frac{\partial y}{\partial x} \sim \left[\frac{F(xi+\Delta xi)-F(xi)}{\Delta xi}\right] \qquad \text{(Eq 9)}$$

$$\frac{\partial y}{\partial x} \sim \left[\frac{(A(xi+\Delta xi)+B)-(A(xi)+B)}{\Delta xi}\right] \qquad \text{(Eq 10)}$$

$$\frac{\partial y}{\partial x} \sim \left[\frac{(A(xi+\Delta xi))-(A(xi))}{\Delta xi}\right] \qquad \text{(Eq 11)}$$

$$\frac{\partial y}{\partial x} \sim \left[\frac{((A\,xi+A\,\Delta xi))-(A(xi))}{\Delta xi}\right] \qquad \text{(Eq 12)}$$

$$\frac{\partial y}{\partial x} \sim \left[\frac{A\,\Delta xi}{\Delta xi}\right] \qquad \text{(Eq 13)}$$

$$\frac{\partial y}{\partial x} = A \qquad \text{(Eq 14)}$$

Note that for a simple linear equation the numerical result for the sensitivity is exactly the same as the analytical expression. There is no restriction on the size of the perturbation (Δxi) used in the calculations. Previously, Freeman [Ref 7] has suggested using a 10% perturbation in xi to do the sensitivity calculations. However any size of perturbation will work and generate the correct sensitivity.

This is a general result. As long as the function can be expressed as a linear function of a number of variables, xi, the sensitivity can be directly calculated by a simple perturbation of the function. The resulting sensitivity will be exactly the same as would be obtained using analytical methods.

Non-Linear Expressions

In the quantification of a fault tree, the top event frequency is computed using failure rates of the component devices and time that measures the interval between system validations. In a minimal cut set of two or more basic events, the time variable may be the same. For example the test interval for a valve may be the same for a pressure transmitter. A minimal cut set involving both the valve and the pressure transmitter would create a 2rd order term in the test interval. If the test interval is uncertain, how do we compute the sensitivity of the top event frequency with respect to the test interval? We will now explore how to deal with this type of problem when completing the uncertainty analysis.

As before, let

$$y = F(xi) = Axi^n + B \qquad \text{(Eq 15)}$$

What is the sensitivity of Y to a change in variable X? By simple calculus:

$$\frac{\partial y}{\partial x} = nAxi^{n-1} \qquad \text{(Eq 16)}$$

Let us evaluate the numerical approximation for this case. If n=2, the function of interest is:

$y = F(xi) = Axi^2 + B$ (Eq 17)

From basic calculus, the exact analytical expression for the sensitivity is:

$$\frac{\partial y}{\partial x} = 2\ A(xi) \quad \text{(Eq 18)}$$

If xi = 1

$$\frac{\partial y}{\partial x} = 2\ A \quad \text{(Eq 19)}$$

Again the approximation of the derivative is:

$$\frac{\partial y}{\partial x} \sim \left[\frac{(A(xi+\Delta xi)^2+B)-(Axi+B)}{\Delta xi}\right] \quad \text{(Eq 20)}$$

$$\frac{\partial y}{\partial x} \sim \left[\frac{F(xi+\Delta xi)-F(xi)}{\Delta xi}\right] \quad \text{(Eq 21)}$$

Inserting the equation for F(xi) we obtain:

$$\frac{\partial y}{\partial x} \sim \left[\frac{(A(xi+\Delta xi)^2+B)-(Axi+B)}{\Delta xi}\right] \quad \text{(Eq 22)}$$

$$\frac{\partial y}{\partial x} \sim \left[\frac{(A\ (xi^2+2\Delta xi+\Delta xi^2)\ +B)-(Axi+B)}{\Delta xi}\right] \quad \text{(Eq 23)}$$

$$\frac{\partial y}{\partial x} \sim \left[\frac{A\ (xi^2+2\Delta xi+\Delta xi^2)\ -Axi}{\Delta xi}\right] \quad \text{(Eq 24)}$$

$$\frac{\partial y}{\partial x} \sim A\left[\frac{(xi^2+2\Delta xi+\Delta xi^2)\ -xi}{\Delta xi}\right] \quad \text{(Eq 25)}$$

Setting xi =1, and using a 10% perturbation as $\Delta xi = 0.1\ xi = 0.1$

$$\frac{\partial y}{\partial x} \sim A\left[\frac{(1^2+(2)(0.1)1+0.1^2\ )\ -1}{0.1\ (1)}\right] \quad \text{(Eq 26)}$$

$$\frac{\partial y}{\partial x} \sim A\left[\frac{(2)(0.1)+0.01}{0.1}\right] \quad \text{(Eq 27)}$$

$$\frac{\partial y}{\partial x} \sim A\left[\frac{0.21}{0.1}\right] \quad \text{(Eq 28)}$$

$$\frac{\partial y}{\partial x} = 2.1\ A \quad \text{(Eq 29)}$$

Note that the numerical estimate (Eq 29) is approximately 5% greater than the analytical value (Eq 19). As shown in Table 1, for higher order equations the error will increase as the size of the perturbation becomes more important. Table 2 shows the impact of reducing the perturbation size to 1%. For 10th order variables (say a minimal cut set with 10 basic events) the error is 5%. For almost all engineering evaluations, a maximum error of 5% should be adequate.

For cases where the test frequency, repair time, or mission time appears in a minimal cut set 3 or more times, I suggest that the following perturbation in time be used to compute the sensitivity of the top event:

$$\Delta time = 0.01\ time \qquad \text{(Eq 30)}$$

The sensitivity of the top event frequency to all uncertain variables can be computed using a perturbation of 1%.

$$\Delta xi = 0.01\ xi \qquad \text{(Eq 31)}$$

3. Fault Tree Analysis

Fault trees are often used to analyze potential failures of an engineered system. A fault tree is a failure logic diagram that shows the relationship of device or system failures leading to an accident. Fault tree analysis is covered in detail in the CCPS CPQRA book [Ref. 1]. A fault tree is constructed by starting with a top event or condition and asking the question:

What conditions lead to this outcome?

Each condition in the next layer down is then analyzed by asking the same question. What conditions lead to this outcome? The logic of how these intermediate conditions (termed intermediate events) are related is expressed using AND and OR logic gates. An AND gate output is true if and only if (iff) all of the inputs to the AND gate are all true simultaneously. An OR gate output is true iff one of the inputs to the OR gate is true.

Once the fault tree logic structure is created, numerical data can be used to calculate the frequency or probability of the top event of concern occurring. For fault trees where a basic event is found in multiple branches of the tree, such as power failure, the logic structure of the tree must be simplified to remove the impact of the repeated events on the top event of concern. This simplification is done using Boolean Algebra. Each resulting grouping of basic events can cause the top event to occur. These groupings of basic events are called minimal cut sets. As a simple example consider the Boolean Algebra representation of a fault tree :

$$T = BE11 + BE1 \bullet BE2 + BE1 \bullet BE3 + BE2 \bullet BE9 + BE3 \bullet BE7 \qquad \text{(Eq 32)}$$

Where:

T = Fault Tree Top Event

BE11 = Basic Event 11

BE1 = Basic Event 1

BE2 = Basic Event 2

BE3 = Basic Event 3

BE7 = Basic Event 7

BE9 = Basic Event 9

+ = Boolean Algebra Addition Symbol

• = Boolean Algebra Multiplication Symbol

For this example the top event T, the minimal cut sets are:

BE11

BE1 BE2

BE1 BE3

BE2 BE9

BE3 BE7

For example the existence of basic events BE1 and BE3 guarantees that the top event T will occur. Publically available standard computer software such as SAPHIRE [Ref. 8] may be used to find the minimal cut sets of a large fault tree.

The representation of the top event in equation 32 can now be used to compute the probability or frequency of occurrence of the top event. If the devices describe by the basic events are non-repairable and are tested after a period of time TI, the probability of top event, T, may be calculated as:

$$\text{Prob (T)} = \lambda_{11}\ TI/2 + (\lambda_1\ TI/2)(\lambda_2\ TI/2) + (\lambda_1\ TI/2)(\lambda_3\ TI/2) + (\lambda_2\ TI/2)(\lambda_9\ TI/2) + (\lambda_3\ TI/2)(\lambda_7\ TI/2) \quad \text{(Eq 33)}$$

Where:

λ_{11} = failure rate of device 11 in basic event BE11, failures per unit time

λ_1 = failure rate of device 1 in basic event BE1, failures per unit time

λ_2 = failure rate of device 2 in basic event BE5, failures per unit time

λ_3 = failure rate of device 3 in basic event BE3, failures per unit time

λ_7 = failure rate of device 7 in basic event BE7, failures per unit time

λ_9 = failure rate of device 9 in basic event BE9, failures per unit time

TI = test interval, time

Equation 33 assumes that the device failure rates are small and the product λ TI < 0.1. The other major assumption is that the values of all of the failure rates and test intervals are known. In reality the failure rate data are not known exactly with only a range or probability distribution representing the state of knowledge.

4. Example Problem

The following example is taken from the ISA technical report on fault tree analysis [Ref. 9]. This example should not be considered as a recommendation by either ISA or AIChE. The example is presented to allow for the demonstration of the methods that can be used in a fault tree study to quantify the uncertainty in the resulting calculated failure frequency or failure probability.

Consider the interlock on the intermediate storage tank T101 shown on Piping and Instrument Diagram in Figure 1. The interlock block diagram is shown in Figure 2. The interlock is intended to prevent an abnormal condition in the tank which could lead to a uncontrolled release of the tank contents. The process hazards analysis team has recommended that the interlock be designed to safety integrity level 2 (SIL 2) with a target probability of failure on demand (PFD) of no greater than 1E-2 or a risk reduction factor (RRF) of 100. Does the proposed interlock shown on Figure 1 meet the SIL 2 target? What is the uncertainty in the predicted PFD of the interlock?

A fault tree for the failure of this interlock is shown in Figure 3. A minimal cut set analysis has been completed and the resulting minimal cut sets are presented in Table 3. We can now write the Boolean Algebra equation that represents this fault tree as:

$$T = PE + TS1 \cdot TS2 + LS1 \cdot LS2 + FT1 \cdot FT2 + FT2 \cdot FT3 + FT1 \cdot FT3 + BV1 \cdot BV2 + BV1 \cdot SOL1 + BV2 \cdot SOL2 + SOL1 \cdot SOL2 + PT1 \cdot PT2 \quad \text{(Eq 34)}$$

Assuming that the interlock is non-repairable until tested at time TI and using lamda-time (λT) approximation of the failure rate, the Boolean representation of the fault tree may now be converted to a failure probability model as:

$$\begin{aligned}\text{Prob}(T) = {} & \text{Prob}(PE) + (\lambda_{TS1}\, TI/2)(\lambda_{TS2}\, TI/2) + (\lambda_{LS1}\, TI/2)(\lambda_{LS2}\, TI/2) + (\lambda_{FT1}\, TI/2)(\lambda_{FT2}\, TI/2) + \\ & (\lambda_{FT2}\, TI/2)(\lambda_{FT3} TI/2) + (\lambda_{FT1}\, TI/2)(\lambda_{FT3}\, TI/2) + (\lambda_{BV1}\, TI/2)(\lambda_{BV2}\, TI/2) + \\ & (\lambda_{BV1}\, TI/2)(\lambda_{SOL1} TI/2) + (\lambda_{BV2}\, TI/2)(\lambda_{SOL2}\, TI/2) + \\ & (\lambda_{SOL1}\, TI/2)(\lambda_{SOL2}\, TI/2) + (\lambda_{PT1}\, TI/2)(\lambda_{PT2}\, TI/2) \qquad \text{(Eq 35)}\end{aligned}$$

Where:

Prob(PE) = failure probability of electronic logic solver, assumed to be constant at 5E-3.

λ_{TS1} = failure rate of temperature switch 1, failures per unit time

λ_{TS2} = failure rate of temperature switch 2, failures per unit time

λ_{LS1} = failure rate of level switch 1, failures per unit time

λ_{LS2} = failure rate of level switch 2, failures per unit time

λ_{FT1} = failure rate of flow transmitter 1, failures per unit time

λ_{FT2} = failure rate of flow transmitter 2, failures per unit time

λ_{FT3} = failure rate of flow transmitter 3, failures per unit time

λ_{BV1} = failure rate of block valve 1, failures per unit time

λ_{BV2} = failure rate of block valve 2, failures per unit time

λ_{SOL1} = failure rate of solenoid valve 1, failures per unit time

λ_{SOL2} = failure rate of solenoid valve 2, failures per unit time

λ_{PT1} = failure rate of pressure transmitter 1, failures per unit time

λ_{PT2} = failure rate of pressure transmitter 2, failures per unit time

TI = test interval, time

The electronic logic solver PE is assumed to have a constant failure probability of 5E-3.

Failure rate data for the other devices are present in Table 4. The mode (most likely value) of the failure rate is set equal to the point value taken from the ISA example [Ref. 9]. The triangular probability distribution (Appendix A) was assumed to represent the failure rate data for the equipment items. The upper and lower limits are based on the failure rate data of Appendix 4 of Smith [Ref. 10].

Using the device failure rates represented by the Table 4 column labeled as the mode and the minimal cut sets presented in Table 3, the ISA technical report [Ref.9, page 35] computes an interlock PFD as 7.5E-3 (RRF of 133). This would satisfy the requirement for a SIL 2 interlock {PFD of 1E-2 or a RRF of 100}. Now what is the uncertainty in the interlock PFD?

The first thing to remember is that the PFD should be calculated based on the mean value of the failure rate lambda (λ), not the mode of the failure rate. For a data set, the mean is the best single point representation of the data set. Table 5 presents the PFD of each minimal cut set using the mean of the device failure rate lambda (λ). The total of the PFD becomes 1.47E-2 (RRF of 68). This would not satisfy a SIL 2 interlock requirement.

Now we will compute the variance of the predicted probability of top event, T. To compute the variance of the probability of the top event, T, we will need to compute the sensitivity of the top event to each of the variables.

Sensitivity Using Numerical Perturbation Calculation

We will use the ISA interlock example previously analyzed by Freeman (Ref. 7). Using public data sources Freeman found that the probability of failure on demand (PFD) was 1.49E-2. To use the perturbation method for the evaluation of the sensitivities requires evaluating the impact of small changes in the uncertain input parameters. To enable these calculations to be completed numbers using 8 significant digits were used. The calculations were completed using Microsoft Excel which uses double-precision floating-point arithmetic compliant with IEEE 754 specification. Excel nominally carries 15 significant figures in calculations. We will complete the calculations with Excel displaying 5 significant digits and round the result at the end of the calculations to 4 significant digits. We will illustrate the calculations by looking at the Flow Transmitter, FT1.

a. First we recalculate the PFD of the example and find it to be 1.4594E-02. This is the base number that we will use in the perturbation calculations.
b. We now make a small change in mean failure rate of FT1. The mean failure rate is 8E-6 failures/hour. A 1% perturbation is used or an increase of 0.08E-6 failures/hour.

c. The perturbed value of the failure rate of FT1 becomes (8 + 0.08) * 1E-6 failures/hour or 8.08E-6 failures/hour.
d. The perturbed value of the failure rate of FT1 is now used to re-calculate the PFD of the interlock. The resulting re-calculation finds that the PFD is 1.4618E-02
e. The change in the PFD from the base number due to the perturbation of the failure rate of FT1 is now computed as:

   delta PFD = 1.4618E-02 - 1.4594E-02 = 2.4556E-05

f. The sensitivity of the PFD to a change in the failure rate of FT1 is now computed by dividing the change in the PFD by the change FT1 failure rate as:

   sensitivity = (2.4556E-05)/(0.08E-6 failures/hour) = 307.0 hours
   Result is rounded to 4 significant digits.

We now note that in the paper by Freeman (Ref. 7) that equation 32 presents the sensitivity for FT1 as:

$$\frac{\partial Prob(T)}{\partial \lambda \mathrm{FT1}} = 307.0 \text{ hr} \qquad \text{(Eq 32 of Ref 7)}$$

The answers are exactly the same. The sensitivity calculations for all of the devices are presented in Table 6. Note that in every case the sensitivity found using the analytical expression and the numerical estimate are the same. This is a general result which will be true for all basic event equipment failure rates found in a fault tree minimal cut set.

Once found, the sensitivities can now be used to compute the variance contribution of each particular device to the overall variance of the top event PFD in the same manner as Freeman previously presented.

The sensitivities of the top event probability are summarized in Table 6. We may now compute the variance contribution of each uncertain variable using Equations 1 – 4 in the methodology section of this paper. The relative variance contribution is expressed as a percent of the total variance of the top event probability. The top event variance is found to be 8.24E-6 (failures per hour)^2 or equivalently the top event standard deviation is 2.78E-3 failures per hour.

In this example the variance (measure of uncertainty) is dominated by the flow transmitters. The flow transmitters account for almost 2/3 of the uncertainty in the top event probability. Failures of the flow transmitters appear in 3 of the minimal cut sets. Since the uncertainty in the transmitter failure rates is large, the flow transmitters should contribute a significant amount to the variance of the fault tree top event.

We can use the variance in the top event to define the likelihood of achieving SIL 2 performance of the interlock. The calculations are done using the normal probability distribution. The normal distribution is tabulated as the standard normal distribution using a normalization factor Z.

The standard normal factor, Z, [Ref. 11] is defined as:

$$Z = \left[\frac{xi - E(x)}{\sigma}\right] \qquad \text{(Eq 36)}$$

Where:

σ = standard deviation. Note that the variance of a random variable is the square of the standard deviation of the random variable.

E(x) = Expected value of the random variable x_i

For the example, the expected value of the probability of the top event of concern was calculated as

E[Prob(T)] =1.47E-2 (Eq 37)

From Table 4, the standard deviation in the probability of the top event is

σ_T = 2.78E-3 (Eq 38)

The target PFD for a SIL 2 interlock must not be worse than 1E-2 (RRF of 100).

We compute the normal distribution Z factor at the SIL 2 target value as

Z = (0.01 – 0.0147)/2.78E-3 (Eq 39)

Z = -1.69065 (Eq 40)

Tabulations of standard normal distribution are presented in most statistics books such as Meyer [Ref. 11]. For this value of Z, the corresponding probability that the interlock will achieve SIL 2 performance is 4.55 percent. We can also compute the risk reduction that the proposed interlock design will achieve.

What is the interlock probability of failure on demand (PFD) that we can be 95% certain that the interlock will provide? At the 95% level, the corresponding Z factor is 1.65 [Ref 11]. We now compute the corresponding PFD of the interlock that corresponds to this Z factor.

$$Z = \left[\frac{x95\% - E[Prob(T)]}{\sigma}\right] \quad \text{(Eq 41)}$$

In this case:

$x_{95\%}$ = the PFD that we can be confident that the interlock will achieve

σ_T =2.78E-3 (Eq 42)

E[Prob(T)] =1.47E-2 (Eq 43)

Z= 1.65 (Eq 44)

Rearranging equation 40 to find $x_{95\%}$we obtain:

$x_{95\%}$ = E[Prob(T)] + Z σ_T (Eq 45)

$x_{95\%}$ = 1.47E-2 + (1.65) (2.78E-3) = 0.0193 (Eq 46)

or a 95% certain risk reduction factor of

RRF = 1/PFD = 1/0.0193 = 52 (Eq 47)

By the same logic there is 5% chance that the risk reduction is a PFD of 0.01 or an RRF of 99. In summary, this interlock is likely to perform as a mid-range SIL 1 interlock, not the SIL 2 interlock desired by process hazards analysis team.

5. Methodology for Fault Tree Uncertainty Analysis

We can now generalize the methods used in the above example to a proposed method for the evaluation of the uncertainty in a fault tree. The following step-by-step procedure is modified from that that previously published by Freeman (Ref 7) to indicate the use of numerical methods to compute the sensitivity. Step 8 shown in *italics* indicates the modification..

1. Create the fault tree using standard methods outline in the CPQRA book [Ref 1].
2. Determine minimal cut sets by hand or using standard computer software such as SAPHIRE [Ref 12].
3. Define the needed failure rate data for each basic event.
4. Define those basic events that are considered to be uncertain.
5. For the basic events that are uncertain, define the probability distribution and associated parameters needed to numerically define the probability distribution. Appendix B of this paper presents the needed description of 4 commonly used probability distributions. In many cases the triangular distribution will be selected. The minimum, maximum and mode (most likely) parameters will be needed for the triangular distribution.
6. Compute the mean of each of the uncertain variable
7. Compute the probability or frequency of the fault tree top event of interest using the mean value for each uncertain variable.
8. *Compute the sensitivity of the top event probability or frequency using the numerical methods outlined in this paper.*
9. Compute the variance of each of the uncertain variables
10. Compute the variance contribution of for each of the uncertain variables to the top event probability or frequency using equation 6.
11. Compute the total variance of the top event of interest probability or frequency by summing all of the contributions determined in Step 10.
12. Compute the variance contribution percent of each uncertain variable by dividing the variable contribution (Step 10) by the total variance of the top event of interest (Step 11).
13. Define the level of risk the project is willing to take. What chance will the project management accept for potential failure of the interlock to achieve the desired risk reduction? In the above example, I have used a 5% risk of failure or a 95% certainty that the interlock will achieve a risk reduction factor of 52. In the example, there is a very low probability that the interlock will achieve a SIL 2 PFD of 0.01 or an RRF of 100.

Alternately you can report the 90% range for the top event of interest. For the example the 90% range starts at the 5% RRF of 99 and there is a 95% certainty that that an RRF of at least 52 will be achieved. The expected risk reduction for this interlock is an RRF of 68.

6. Conclusions

A general procedure has been presented to quantify the uncertainty in calculations of failure frequencies using the fault tree methodology. The uncertainty analysis procedure is based on the application of propagation of error and variance contribution analysis techniques to the minimal cut sets created during a fault tree study. The procedure is simple and can be incorporated into standard fault tree analysis programs such as SAPHIRE [Ref 8]. This procedure is based on perturbation calculation of the frequency of the top event from changes in the uncertain variables. Incorporation of a numerical perturbation method into standard fault tree software such as SAPHIRE would be easy to implement and would provide the means for evaluation of the uncertainty in the quantified fault tree results.

For the example interlock presented in this paper, the uncertainty in the device failure rates degrades the predicted interlock SIL from meeting SIL 2 requirements to becoming a mid range SIL 1 interlock. This is a general result. Uncertainty in design parameters will reduce the likelihood that the equipment will achieve desired results. The design engineer has two choices:

- Accept the conclusion that the interlock will not perform as SIL 2
- Re-design the interlock using more reliable components to achieve the SIL 2 target

The variance contribution analysis shown in Table 7 indicates that the major contributor to the uncertainty in the interlock performance is the Flow Transmitters. Flow transmitters certified to SIL-2 or SIL-3 performance with lower failure rates could improve the calculated PFD of the interlock.

7. References

1. Center for Chemical Process Safety , *Guidelines for Chemical Process Quantitative Risk Analysis – 2rd edition*, American Institute of Chemical Engineers, New York, 2000.
2. Frank P Lees, Loss Prevention in the Process Industries – 2rd Edition, Butterworth, ISBN 0-7506-1547-8, Oxford, UK, 1996
3. US Department of Labor, Occupational Safety and Health Administration, CFR 1910.119 - Process Safety Management of Highly Hazardous Chemicals, 1910.119(e)(2)(vi).
4. Edward B. Haugen, *Probabilistic Approaches to Design*, John Wiley, New York, 1968
5. Raymond. Freeman and Angela Summers, "Evaluation of Uncertainty in Safety Integrity Level Calculations," *Process Safety Progress,* Published online in Wiley Online Library (wileyonlinelibrary.com) DOI 10.1002/prs. 11805, 2016
6. Raymond. Freeman, "General Method for Uncertainty Evaluation of Safety Integrity Level Calculations – Part 2 Analytical Methods," *Process Safety Process*, Published online in Wiley Online Library (wileyonlinelibrary.com) DOI 10.1002/prs 111915, October 2017
7. Raymond. Freeman ,"Error Propagation and Uncertainty Analysis: Application to Fault Tree Analysis", Process Safety Progress, Published Wiley Online Library (wileyonlinelibrary.com/journal/prs) DOI: https://doi.org/10.1002/prs.12080, June 29, 2019

8. US Nuclear Regulatory Commission, "Systems Analysis Programs for Hands-on Integrated Reliability Evaluations (SAPHIRE) – Version 8," NUREG/CR-7039, Vol 3, June 2011.
9. ISA, "Safety Instrumented Functions (SIF) – Safety Integrity Level (SIL) Evaluation Techniques Part 3: Determining the SIL of a SIF via Fault Tree Analysis, ISA Technical Report ISA-TR-84.00.02-2002 – Part 3, Approved 17 June 2002.
10. David J. Smith, *Reliability Maintainability and Risk – 8th Edition*, ISBN 978-0-08-096902-2, Elsevier, London, 2011
11. Paul L. Meyer, Introductory Probability and Statistical Applications, John Wiley, New York, 1972

Table 1. Comparison of Analytical Sensitivity with Numerical Approximation with Perturbation of 10% in xi for the equation: y = F(xi) = $Axi^n + B$

Order of Equation n	Equation for Analytical Sensitivity $\frac{\partial y}{\partial x} = nAxi^{n-1}$	Sensitivity with xi = 1 and A=1	Numerical Sensitivity with $\Delta xi = 0.1\, xi$	Error %
1	$\frac{\partial y}{\partial x} = 1A(xi)^0$	1	1	0
2	$\frac{\partial y}{\partial x} = 2A(xi)^1$	2	2.1	5
3	$\frac{\partial y}{\partial x} = 3A(xi)^2$	3	3.31	-10%
4	$\frac{\partial y}{\partial x} = 4A(xi)^3$	4	4.64	-16%
5	$\frac{\partial y}{\partial x} = 5A(xi)^4$	5	6.11	-22%
6	$\frac{\partial y}{\partial x} = 6A(xi)^5$	6	7.72	-29%
7	$\frac{\partial y}{\partial x} = 7A(xi)^6$	7	9.49	-36%
8	$\frac{\partial y}{\partial x} = 8A(xi)^7$	8	11.44	-43%
9	$\frac{\partial y}{\partial x} = 9A(xi)^8$	9	13.58	-51%
10	$\frac{\partial y}{\partial x} = 10A(xi)^9$	10	15.94	-59%

Error = (Analytical- Numerical)/Analytical

Table 2. Comparison of Analytical Sensitivity with Numerical Approximation with Perturbation of 1% in xi for the equation: $y = F(xi) = Axi^n + B$

Order of Equation n	Equation for Analytical Sensitivity $\frac{\partial y}{\partial x} = nAxi^{n-1}$	Sensitivity with xi = 1 and A=1	Numerical Sensitivity with $\Delta xi = 0.1\, xi$	Error %
1	$\frac{\partial y}{\partial x} = 1A(xi)^0$	1	1	0
2	$\frac{\partial y}{\partial x} = 2A(xi)^1$	2	2.01	-1%
3	$\frac{\partial y}{\partial x} = 3A(xi)^2$	3	3.03	-1%
4	$\frac{\partial y}{\partial x} = 4A(xi)^3$	4	4.06	-2%
5	$\frac{\partial y}{\partial x} = 5A(xi)^4$	5	5.10	-2%
6	$\frac{\partial y}{\partial x} = 6A(xi)^5$	6	6.15	-3%
7	$\frac{\partial y}{\partial x} = 7A(xi)^6$	7	7.21	-3%
8	$\frac{\partial y}{\partial x} = 8A(xi)^7$	8	8.29	-4%
9	$\frac{\partial y}{\partial x} = 9A(xi)^8$	9	9.37	-4%
10	$\frac{\partial y}{\partial x} = 10A(xi)^9$	10	10.46	-5%

Error = (Analytical- Numerical)/Analytical

Table 3. Basic Events in Minimal Cut Sets For Example

Cut Set No	BE 1	BE 2
1	PE	-
2	TS1	TS2
3	LS1	LS2
4*	FT1	FT2
5*	FT2	FT3
6*	FT1	FT3
7*	BV1	BV2
8*	BV1	SOL1
9*	BV2	SOL2
10*	SOL1	SOL2
11	PT1	PT2

* Minimal cut Set with same basic event as another cut set

Table 4. Lamda (λ) Failure Rate Data With Uncertainty Limits

Device	Minimum Failures per 1E6 Hours	Mode Failures per 1E6 Hours	Maximum Failures per 1E6 Hours	Mean Failures per 1E6 Hours	Lamda Variance (Failures per 1E6 Hours)^2
Flow Transmitters FT1, FT2 and FT3	1.0	2.9	20.0	8.0	18.3
Pressure Transmitters PT1 and PT2	1.0	2.3	20.0	7.8	18.8
Temperature Switch TS1 and TS2	3.0	7.6	20.0	10.2	12.9
Level Switch LS1 and LS2	2.0	4.6	20.0	8.9	15.8
Block Valves BV1 and BV2	0.2	2.3	10.0	4.2	4.4
Solenoid Valves SOL1 and SOL2	1.0	2.3	8.0	3.8	2.3

Table 5. Probability of Minimal Cut Sets Based on Mean Failure Rates

Cut Set No	BE 1	BE 2	PFD based on mean λ
1	PE	-	5.00E-03
2	TS1	TS2	2.00E-03
3	LS1	LS2	1.52E-03
4	FT1	FT2	1.23E-03
5	FT2	FT3	1.23E-03
6	FT1	FT3	1.23E-03
7	BV1	BV2	3.38E-04
8	BV1	SOL1	3.38E-04
9	BV2	SOL2	3.38E-04
10	SOL1	SOL2	2.77E-04
11	PT1	PT2	1.17E-03

Table 6. Sensitivity of Example System PFD to Basic Event Failure Rates Based on a 1% Perturbation

Basic Event Label	Device Type	Mean Failures per 1E6 Hours	Failure Perturbation of 1% of Mean Failures per 1E6 Hours	Perturbed Failure Rate Failures per 1E6 Hours	Total System PFD Using Perturbed Failure Rate of Basic Event	Change in Total System PFD	Sensitivity of PFD to Change in Failure Rate Hours
FT1	Flow Transmitter	8	0.08	8.08	1.461848E-02	2.455603E-05	307.0
FT2	Flow Transmitter	8	0.08	8.08	1.461848E-02	2.455603E-05	307.0
FT3	Flow Transmitter	8	0.08	8.08	1.461848E-02	2.455603E-05	307.0
PT1	Pressure Transmitter	7.8	0.078	7.878	1.460560E-02	1.167179E-05	149.6
PT2	Pressure Transmitter	7.8	0.078	7.878	1.460560E-02	1.167179E-05	149.6
TS1	Temperature Switch	10.2	0.102	10.302	1.461389E-02	1.995945E-05	195.7
TS2	Temperature Switch	10.2	0.102	10.302	1.461389E-02	1.995945E-05	195.7
LS1	Level Switch	8.9	0.089	8.989	1.460912E-02	1.519596E-05	170.7
LS2	Level Switch	8.9	0.089	8.989	1.460912E-02	1.519596E-05	170.7
BV1	Block Valves	4.2	0.042	4.242	1.460037E-02	6.445958E-06	153.5
BV2	Block Valves	4.2	0.042	4.242	1.460037E-02	6.445958E-06	153.5
SOL1	Solenoid Valves	3.8	0.038	3.838	1.459976E-02	5.832058E-06	153.5
SOL2	Solenoid Valves	3.8	0.038	3.838	1.459976E-02	5.832058E-06	153.5

Table 7. Variance of Top Event Probability Due to Uncertain Device Failure Rates

Device No	Device Type	Label	Sensitivity of Top Event Probability to Device Failure Rate $Hour^{-1}$	Variance of Device Failure Rate (Failures per 1E6 $Hour)^2$	Variance Contribution to Top Event Variance	Contribution to Top Event Variance %
1	Logic Solver	PE	-	0	0.00E+00	0%
2	Temperature Switch	TS1	195.7	12.9	4.94E-07	6%
3	Temperature Switch	TS2	195.7	12.9	4.94E-07	6%
4	Level Switch	LS1	170.7	15.8	4.60E-07	6%
5	Level Switch	LS2	170.7	15.8	4.60E-07	6%
6	Flow Transmitter	FT1	307	18.3	1.72E-06	21%
7	Flow Transmitter	FT2	307	18.3	1.72E-06	21%
8	Flow Transmitter	FT3	307	18.3	1.72E-06	21%
9	Block Valve	BV1	153.5	4.4	1.04E-07	1%
10	Block Valve	BV2	153.5	4.4	1.04E-07	1%
11	Pressure Transmitter	PT1	149.6	18.8	4.21E-07	5%
12	Pressure Transmitter	PT2	149.6	18.8	4.21E-07	5%
13	Solenoid valve	SOL1	153.5	2.3	5.42E-08	1%
14	Solenoid valve	SOL2	153.5	2.3	5.42E-08	1%

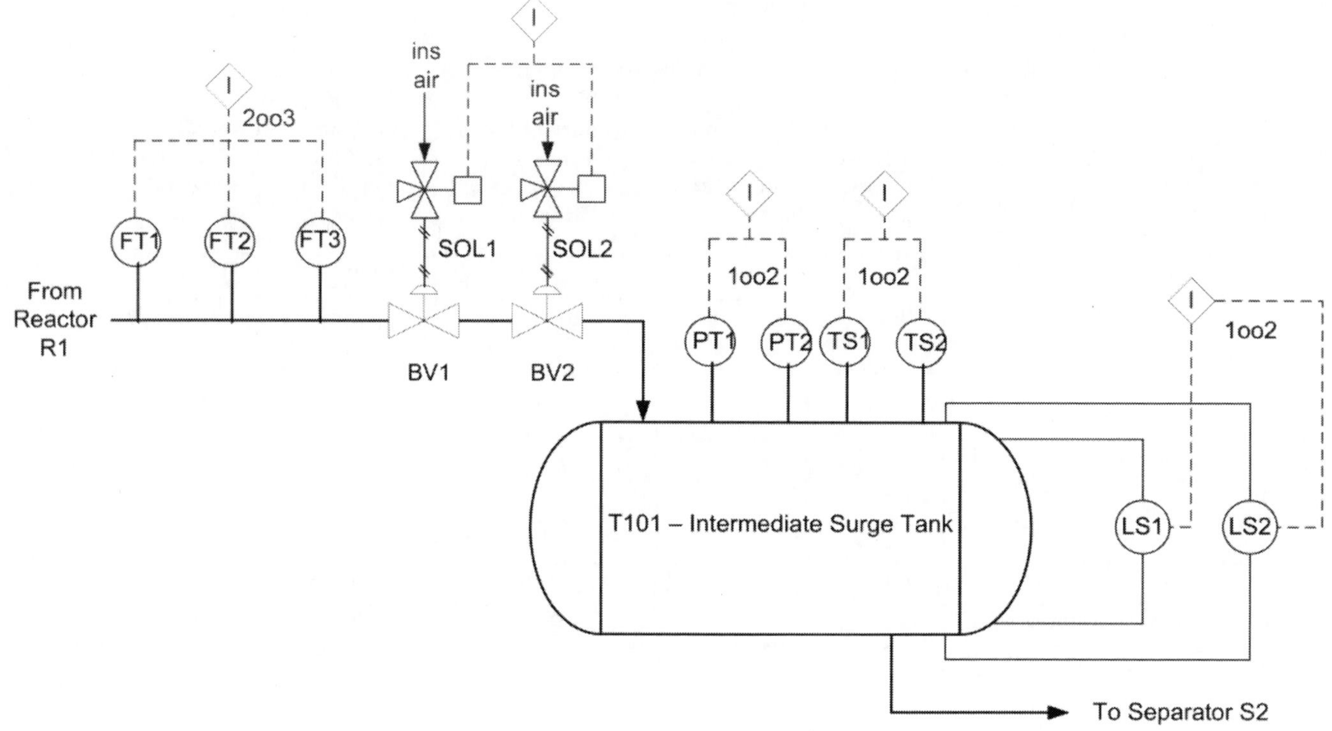

Figure 1. P&ID Diagram for Example

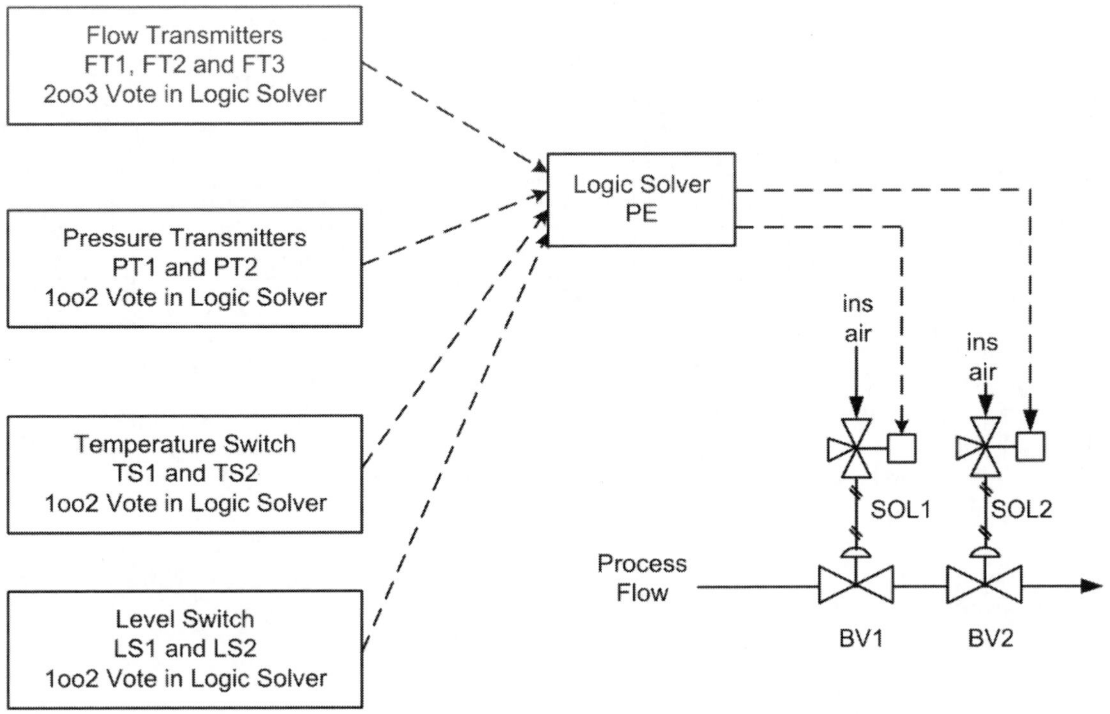

Figure 2. Interlock Block Diagram

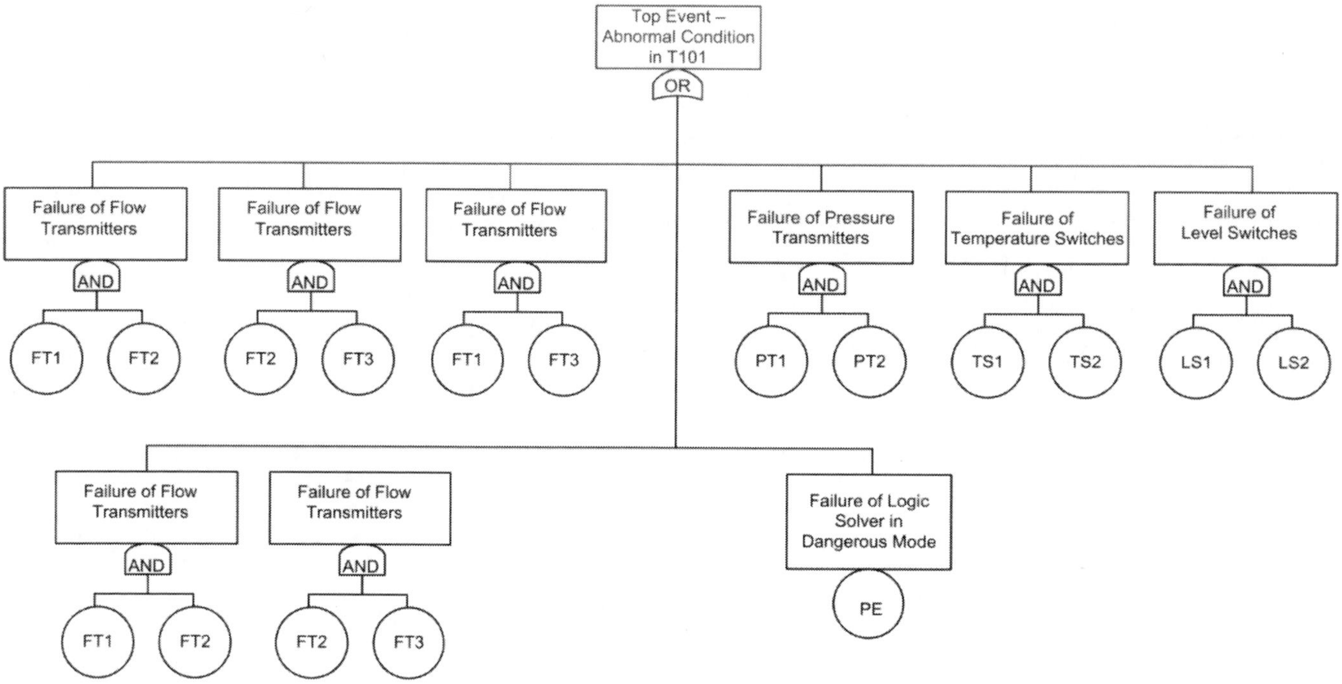

Figure 3. Fault Tree for Example Interlock

Appendix A – Commonly Encounter Probability Distributions

There are four probability distributions used to represent the failure rate of equipment items. These are:

1. Uniform Distribution
2. Triangular Distribution
3. Normal Distribution
4. Log-Normal Distribution

This appendix reviews the properties of these four probability distributions.

1. Uniform Distribution

The uniform distribution [Ref. A-1, pg 74; A-2, pg 687-688] is used when a variable is only known within a defined range. Figure A-1 presents a plot of the probability density function (pdf) for the uniform distribution. The distribution is defined using two parameters, A and B.

Where:

A = minimum value of the random variable

B = maximum value of the random variable

The mean or expected value of a random variable, x, that follows the uniform distribution is:

Mean = Expected Value – $E(x) = (A+B)/2$ (Eq A-1)

The variance of a random variable, x, that follows the uniform distribution is:

Variance = $V(x) = (B-A)^2/12$ (Eq A-2)

The standard deviation of a random variable, x, that follows the uniform distribution is:

Standard Deviation = $(\text{Variance})^{1/2} = (B-A)/(12)^{1/2}$ (Eq A-3)

2. Triangular Distribution

The triangular distribution [Ref. A-2, pg 686-687] is used when a variable is known to lie within a defined range and a "best guess" or estimate can be made as to the most likely value to the variable. Figure A-2 presents a plot of the probability density function (pdf) for the triangular distribution. The distribution is defined using three parameters, A. B and C.

Where:

A = minimum value of the random variable

B = maximum value of the random variable

C = most likely (mode) of the random variable

The mean or expected value of a random variable, x, that follows the triangular distribution is:

Mean = Expected Value – $E(x) = (A+B+C)/3$ (Eq A-4)

The variance of a random variable, x, that follows the triangular distribution is:

Variance = V(x) = $[A^2 + B^2 + C^2 - AB - AC - BC]/18$ (Eq A-5)

The standard deviation of a random variable, x, that follows the triangular distribution is:

Standard Deviation = $(\text{Variance})^{1/2} = \{[A^2 + B^2 + C^2 - AB - AC - BC]/18\}^{1/2}$ (Eq A-6)

3. Normal distribution

The normal distribution [Ref A-2, pg 665-666] is used when a variable when data analysis finds the normal distribution is the best model to describe the spread in the measured variable. Figure A-3 presents a plot of the probability density function (pdf) for the normal distribution. The distribution is defined using two parameters, μ and σ.

Where:

μ = mean of the measured data for variable

σ = standard deviation of the measured data for variable

The expected value of a random variable, x, that follows the normal distribution is:

Expected value = mean = $E(x) = \mu$ (Eq A-7)

The variance of a random variable, x, that follows the normal distribution is:

Variance = square of standard deviation = $V(x) = \sigma^2$ (Eq A-8)

The standard deviation of a random variable, x, that follows the normal distribution is:

Standard deviation = σ (Eq A-9)

4. Lognormal Distribution

The lognormal distribution [Ref. A- 2, pg 658-659] is used when a variable when data analysis finds the lognormal distribution is the best model to describe the spread in the measured variable. Figure A-4 presents a plot of the probability density function (pdf) for the lognormal distribution. Many variables in nature are lognormally distributed (for example the height of people). The log normal distribution is used when the variable of interest has a known physical lower limit of zero. A variable, x, is lognormally distributed if ln(x) is normally distributed. The lognormal distribution is defined using two parameters, μ_y and σ_y.

Let the original data be defined by the variable x with mean and standard deviation as:

μ_x = mean of the measured data for variable x

σ_x = standard deviation of the measured data for variable x

Define a new variable y as:

$y = \text{Ln}(x)$ (Eq A-10)

Then variable x is said to be lognormally distributed if y is normally distributed. We can now write the mean and standard deviation of y (μ_y and σ_y.) [Ref. A-3, A-4 and A-5] as

$$\mu_y = Ln\left[\mu_x^2 /((\sigma_x^2 + \mu_x^2)^{1/2})\right] \quad \text{(Eq A-11)}$$

$$\sigma_y = \left[Ln(\sigma_x^2 / \mu_x^2 + 1)\right]^{1/2} \quad \text{(Eq A-12)}$$

Where:

Ln is the natural logarithm (base e) of the argument

Appendix A References:

A-1. P. L. Meyer, *Introductory Probability and Statistical Applications*, 2nd Edition, Addison-Wesley, Reading Mass, Library of Congress Catalog no. 75-104971, 1972, pp. 74

A-2. D. Vose, *Risk Analysis–A Quantitative Guide*, 3rd edition, Wiley, Chichester, West Sussex, England, ISBN 978-0-470-51284-5, 2008, pp. 686-687

A-3. "Lognormal mean and variance - MATLAB lognstat". www.mathworks.com. Retrieved 27 December 2018

A-4. M. Mood, F. A. Graybill, and D. C. Boes. *Introduction to the Theory of Statistics.* 3rd ed., New York: McGraw-Hill, ISBN 0-07-042864-6, 1974. pp. 117.

A-5. E.L. Crow and K. Shimizu, Lognormal Distributions – Theory and Applications, Marcel Dekker, New York, ISBN 0-8247-7803-0, 1988, pp. 9

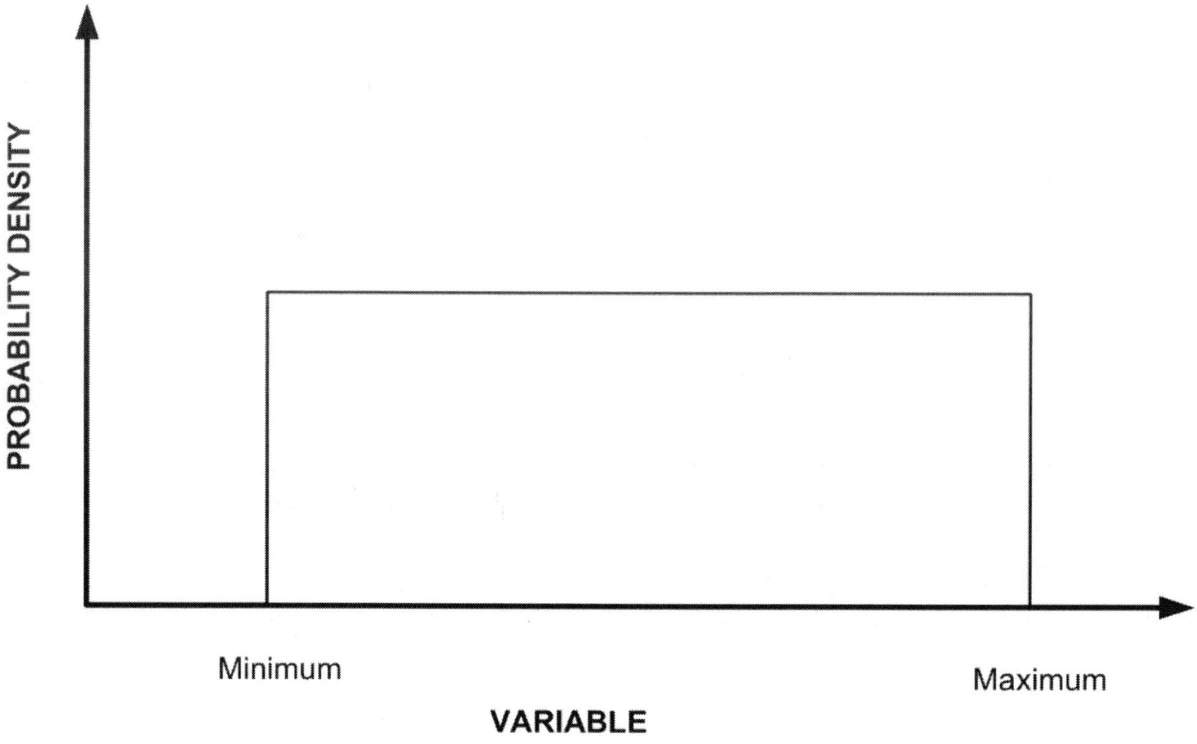

Figure A-1. Uniform Probability Distribution

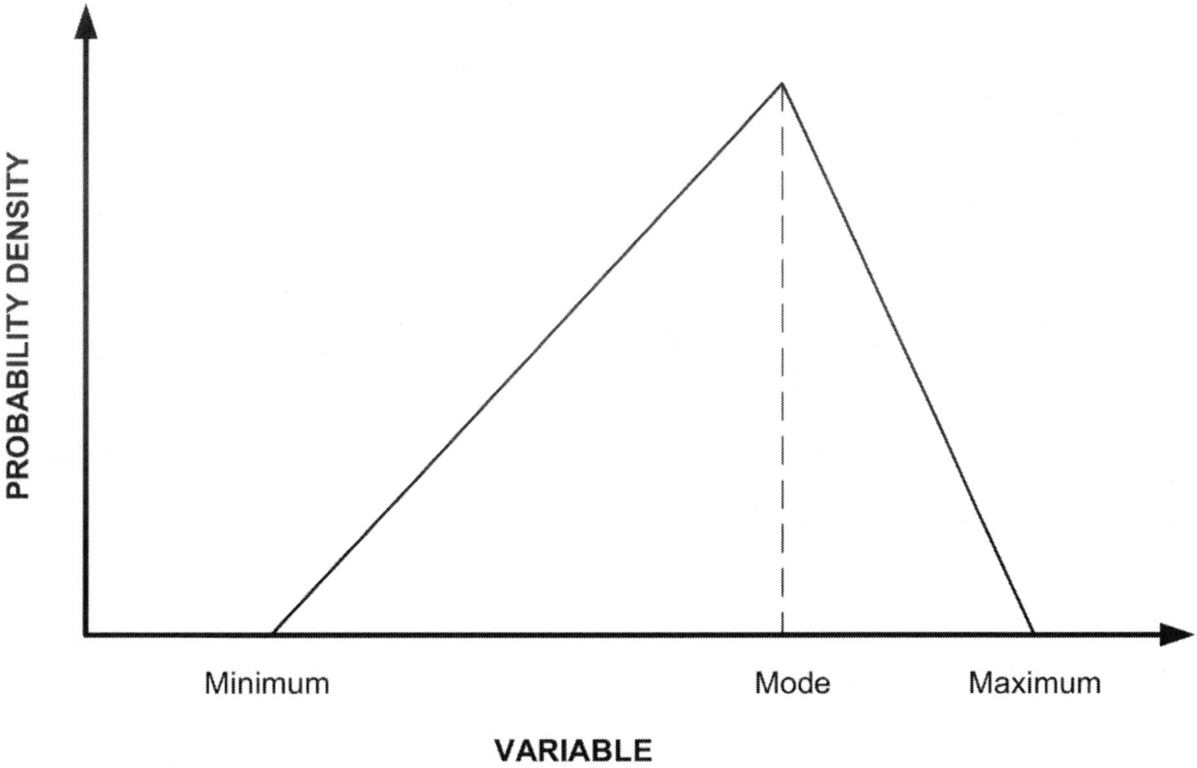

Figure A-2. Triangular Probability Distribution

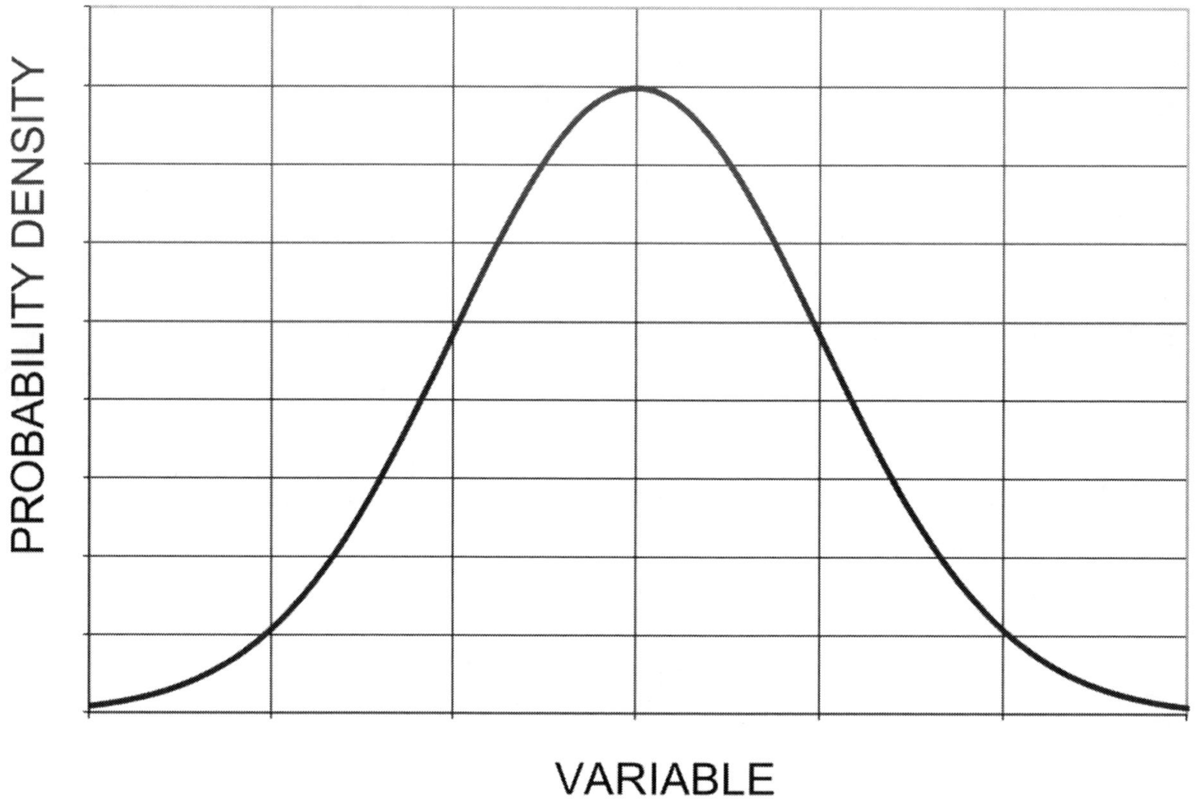

Figure A-3. Normal Probability Distribution

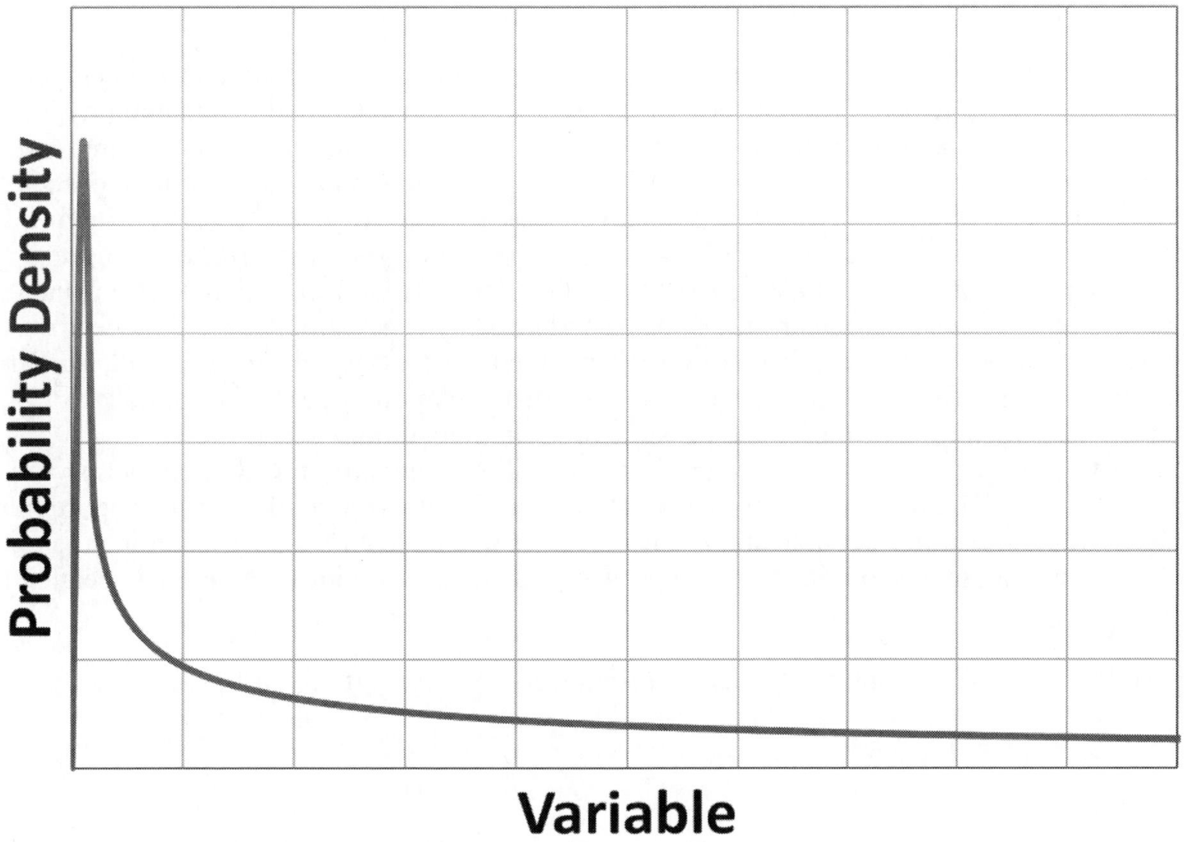

Figure A-4. Lognormal Probability Distribution

590424 Making the Most of Your PHA

Abstract:

A Process Hazard Analysis (PHA) is required to address OSHA Process Safety Management (PSM) and EPA Risk Management Program (RMP) regulations. Understanding the potential process hazard is widely recognized as a foundational element of PSM and RMP, and in addition to simply meeting regulatory requirements, can be a critical mechanism for flushing out potential vulnerabilities early in the design process for capital projects. Although an important activity, PHA is also recognized as a sizeable investment in time and energy for an Owner/Operator. So why not make the most of it? Other important activities at plants such as Mechanical Integrity, Alarm Rationalization, and Management of Change are able to stay focused and possibly be streamlined (i.e., reduced cost) based on the ability to use the results of a quality PHA having the necessary information and usability of documentation. So why not structure the PHA and documentation to be able to capitalize on potentially significant cost-savings and possible program integration?
The thrust of this paper and conference presentation will be to identify tips for conducting and documenting the PHA to maximize its ability to be used for supporting other important programs such as MOC, MI, and Alarm Rationalization. The objective will be to allow PHA Coordinators to be able to readily apply these tips to bring about improvements and cost-savings on these interrelated programs.

589814 Theoretical Analysis of the Viability of Isopropanol-Cyclohexane Solvent Systems for Biomatter Lipid Extraction

Abstract:

A theoretical alternative solvent combination for the extraction of lipids from biomatter is the Water, Isopropanol, and Cyclohexane instead of the more common combination used in the Bligh and Dyer Method (Water, Methanol, and Chloroform). While the Bligh and Dyer Method is considered effective, toxicological concerns have been raised over its use of both Chloroform and Methanol. The alternative method utilizes solvents that do not pose as high as a toxicological concern while theoretically maintaining a similar level of practical viability. The alternative method's extraction efficacy was judged against the performance of the Bligh and Dyer method using the simulation software Aspen Plus V10®. The equilibrium data generated from Aspen Plus V10® was then graphed into tertiary diagrams for data analysis using the graphing software SigmaPlot.

592166 Techno-Economic Analysis of the Conversion of CH4 and CO2 Intoacetic Acid Using Microwave Plasma

Abstract:

Methane and carbon dioxide together represent 90% of greenhouse gases [1], and one way of diminishing their emissions is through the development of pathways that convert them into added value chemicals. Acetic acid is one of such chemicals, and efforts at the laboratory scale have been carried out to obtain alternative ways of producing it. However, there are very few studies that evaluate the possibility of implementing these processes at a larger scale. In this paper, techno-economic and sustainability analyses were used to assess the feasibility of upscaling two pathways to produce acetic acid: (1) by direct conversion of methane and carbon dioxide and (2) by indirect conversion, using syngas (obtained from dry methane reforming) as an intermediate. For the direct method a microwave non-catalytic plasma reactor was considered as the main driver in the process simulation [2]. For the indirect method, a zeolite catalytical single step reaction unit was considered [3]. APEN Plus was used to simulate the reaction and purification systems of both methods, and alongside literature values and plant cost estimation techniques, the capital and operational expenses were calculated. Finally, the feasibility of both projects is assessed on carbon sequestration mass balances and economic metrics. The main sensitivity variables that need to be improved in order to obtain a minimum ROI threshold value were also evaluated.

Key words: carbon management, process design, microwave plasma reactor

[1] EPA. *Overview of Greenhouse Gases*. United States Environmental Protection Agency. 2017.

[2] Wang, L., Yi, Y., Wu, C., Guo, H., Tu, X. *One-Step Reforming of CO2 and CH4 into High-Value Liquid Chemicals and Fuels at Room Temperature by Plasma-Driven Catalysis*. Angewandte Chemie. 2017.

[3] Lopez, L., Velasco, J., Montes, V., Marinas, A., Cabrera, S., Boutonnet, M., Jaras, S. *Synthesis of Ethanol from Syngas over Rh/MCM-41 Catalyst: Effect of Water on Product Selectivity*. Catalysts. 2015.

589668 Model Predictive Control Design for the Allam Power Plant Integrated with an Air Separation Unit

Abstract:

Since electricity and heat generation account for around 25% of this greenhouse gas emission, high efficiency power generation technology such as the Allam cycle which emits essentially no CO_2 and NOx is much needed. Being an oxy-fuel power cycle, the Allam power plant is inherently integrated with an air separation plant (ASU). In this study Aspen Plus Dynamics was used to model the integrated Allam power plant/Air Separation Unit (ASU) with a high degree of heat and work integration based on the steady state work of Net Power. In our previous work, regulatory and advanced PID controls were implemented and tested for various set point tracking and disturbance rejection scenarios. However, PID controllers have limitations for various reasons. Consequently, a model predictive control (MPC) was designed in this study to optimize and improve the performance of the integrated unit and to test its performance in terms of dynamic behavior and robustness to model uncertainties.

589672 Explore the Function of Ion-Exchange Membrane (IEM) in Membrane Capacitive Deionization (mCDI) Cell, an Alternative Technique in Brackish Water Desalination Field

Abstract:

In the arid west, the freshwater supply of many communities is limited, leading to increased interest in tapping brackish water resources. Although reverse osmosis is the most common technology to upgrade saline waters, there is also interest in developing and improving alternative technologies. Here we focus on Capacitive Deionization (CDI), which has attracted broad attention as an energy efficient desalination technology, especially for ion extraction from low salinity brackish water. The objective of this work is to explore the function of ion-exchange membrane (IEM) in enhancing the performance of CDI cell. In this study, a fully coupled two-dimensional model describing ion transport and adsorption behavior in Membrane Capacitive Deionization (mCDI) cell was developed and solved numerically, taking advantage of Darcy's Law, Nernst-Planck equation, and Donnan equilibrium considering hydraulic dispersion in channel, non-ideal membrane, and modification of ion diffusivity in porous media. The model behavior was supported by the experimentally observed variations in transient effluent concentration under different operating conditions. The results showed that ion flux was significantly improved because of the IEM, leading to a faster desalination rate. The equilibrium adsorption amount is increased due to the permselectivity of the IEM, resulting in extra ion accumulation in electrode's macropores during desalination process. The cut-off time (optimal time to switch polarity to generate highest energy efficiency) of mCDI was shorter with a higher salt removal percentage compared to CDI under the same operating conditions. A sensitivity study of the thickness and the fixed charge density (starting from 500mM) of the IEM was done separately, and neither of them were found to be the dominant influencing factors under a low concentration range of brackish water (below 100mM) in this study.

590103 An Integrated Process for the Production of Propylene Glycol from Shale Gas

Abstract:

Propylene glycol (PG) is produced on a large scale primarily for the production of environmentally safest and least toxic industrial and domestic products such as polyester resins, thermoset plastics, polyurethanes, food additives and antifreeze. In this research, a novel integrated plant is developed for the production of PG from shale gas. This integrated approach has the benefit of safer operating conditions since the intermediate propylene oxide, which is explosive, does not need to be stored and transported. Furthermore, there are potential economic benefits from integration. The overall plant is simulated in the Aspenplus environment, and a variety of process conditions are tested at steady-state to optimize the production of PG. Heat-integration tools are utilized for energy-saving and capital cost reduction opportunities. A comparative economic assessment based on existing plant information indicates that the use of process integration techniques has the potential to reduce costs significantly.

Keywords: Process Integration, Heat Integration, Propylene Glycol

592456 Monitoring Methods for Fault Detection and Diagnosis: Algorithms and Applications

Abstract:

Many industrial processes collect an abundance of data from different sensors. These sensors often measure a wide variety of physical properties in order to ensure that these parameters are adhering to expected values. This is essential to ensure plant and operator safety, increase economic benefits, and maintain product quality.

A number of algorithms have been developed in order to improve existing fault detection and diagnosis performance. These algorithms integrate a number of different data-driven driven tools and methods. Multiscale wavelet-based representation of data can be used in order to handle data that is autocorrelated, non-Gaussian, and noisy. Hypothesis testing methods such as the Generalized Likelihood Ratio (GLR) technique can be used in order to provide the best possible detection for a fixed false alarm rate. Moreover, certain model-based methods have been developed in order to monitor process drifts and degradations in the process model, even when a process is operating under control.

This work will discuss the different tools utilized in development of the algorithms, their features, and practical applications, including the benchmark Tennessee Eastman Process (TEP).

589902 Selective Separation of Industrial Relevant Metal Ions from High Tds Water Using Wafer-Enhanced Electrodeionization (WE-EDI)

Abstract:

There is a need for an efficient and economical process of ion-selective separation. In hydraulic fracturing, for instance, several million gallons of water are used in a site, most of which are disposed of by deep well injection taking the water from the ecosystem and causing earthquakes. However, due to the high amount of ions in wells at fracking sites, pressure driven processes for water recovery are impossible to use. This project will demonstrate an emerging environmentally friendly Wafer Enhanced Electrodeionization (WE-EDI) technology which attempts to eliminate the shortcomings that come with normal EDI and separates effluent streams into concentrated ions and clean water. In WE-EDI, ion-exchange resins are replaced by wafers that consist of cation and anion exchange beads and a polymer binding agent. In this study, selective separation of these ion exchange wafers and the effect of electroosmotic drag on the membrane structure have been performed for various anions and cations (Na+, K+, Ca2+, Cl-) and improved selectivity and higher water recoveries have been observed while significant modeling has taken place to predict the selective separation as a strong function of the equilibrium constants between the ions and the beads. This project is supported by the Membrane Science and Technology (MAST) Center and is conducted with the collaboration and guidance of industry mentors.

593303 Influence of Particles on Amine Losses during CO_2 Capture: A Process Simulation Coupled Aerosol Dynamics Model _

Abstract:

Influence of Particles on Amine Losses during CO_2 Capture: A Process Simulation Coupled Aerosol Dynamics Model

David Dhanraj and Pratim Biswas

Aerosol and Air Quality Research Laboratory

Department of Energy, Environmental and Chemical Engineering

Washington University in St. Louis

St. Louis, MO 63130, USA

Efforts to reduce the concentration of atmospheric CO_2 have been initiated due to escalating concerns of global warming and climate change. Amine-based post-combustion CO_2 capture offers an economically viable retrofit for power-plants in combating these emissions. However, solvent losses have been a major challenge to commercializing this technology. Recent experiments[1] show that vapor-based and aerosol-based solvent losses are the key mechanisms of amine solvent loss during CO_2 capture. While vapor-based amine losses can be abated by employing techniques like water-washing, there has been little work to curtail aerosol-based solvent losses. This is primarily because the underlying mechanisms governing these losses are still not well understood. Although there are a few numerical models proposed for this system, there are no models that consider aerosol phenomena such as coagulation or particle loss. Moreover, these models have not been fully validated through experiments.

Furthermore, there are a few techniques that are being developed to tackle solvent losses. It is known that reducing particle number concentration in the flue gas will reduce amine emissions. Conventional Electrostatic Precipitators (ESP) are however inefficient in capturing particles in the size range 0.1 – 1.0 μm. Amongst different approaches demonstrated, introducing a pre-charging stage and a photo-ionization stage has shown potential in increasing the charging efficiency.

Therefore, the broad objectives of this work are primarily focused on two facets of the amine loss problem. The first is the development of a model that integrates aerosol phenomena with a process engineering model that accounts for the reactive mass transfer process in a packed-bed absorber. The second is the design, fabrication and testing of a bench-scale Photoionization Enhanced Two-staged Electrostatic Precipitator (PI-ESP) at Abbott Power Plant. In essence, with the PI-ESP aimed at reducing the particulate concentration in the flue gas, and the model relating the influence of particulate concentration in the flue gas to amine emissions, this work aims to illuminate the underlying mechanisms governing amine emissions, and demonstrate a solution to tackle the problem.

References

1. International Journal of Greenhouse Gas Control, vol. 85, pp. 86-99, 2019.
2. International Journal of Greenhouse Gas Control, vol. 19, pp. 138-144, 2013.

589417 Yay! I'm a Manager...Now What?

Abstract:

You've just been promoted to your first supervisory role, but the landscape feels strange. You are spending more time with people than with technical issues. It seems like there are no "right" answers and you thought you'd be working less hours with more recognition!

The reality is that managers need different skills to be successful than we've learned in our traditional chemical engineering education. As managers, we need to learn delegation, negotiation, and conflict and time management to be successful. In this presentation, you will learn tips and techniques to identify the severity of team conflicts and how to utilize creative conflict to improve team performance.

If you're a new or seasoned manager or faculty member, you will want to attend this presentation to learn what drives conflict and how to build trust on your teams for increasing innovation. These skills are vital for new managers transitioning from a technical role to leading people.

591461 Successfully Navigating Generational Differences

Abstract:

//

Generational Differences

- Perspectives from the Younger Generation
  - by Lori McDowell, Director, Business Development, Matheson

A Note About this Presentation

- Many presentations have been given on generational differences and most have focused on how the younger generations don't meet the expectations of the workplace
- This session was planned as a panel of young engineers to discuss generational differences from their perspective
- Since we can't have a live panel, I have spoken to many young engineers and will be using many of their comments as the basis for this presentation

Outline

- The Stereotypes about different Generations in the Workforce
- The Truth about different Generations in the Workforce
- Comments from Successful Young Engineers
- Tips for how we can all just get along
- Thank you to contributors

A Little Humor about Generations

- The Job Interview
  - https://www.youtube.com/watch?v=Uo0KjdDJr1c
  - https://www.youtube.com/watch?v=Ed-5Zzdbx0E

Five Generations Working Side by Side in 2020

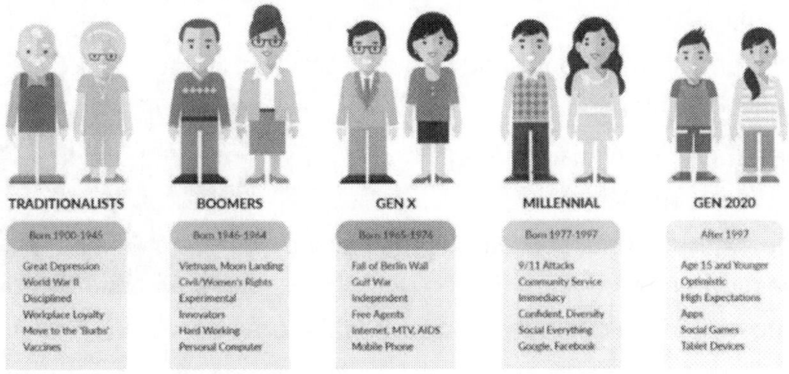

Generational Stereotypes

- Traditionalists
  - Patriotism, sacrifice
  - Teamwork
  - Follow directions and respect authority
- Baby Boomers
  - Live to work, their career is their persona
  - Wealthiest generation, consumers, concerned about retirement
  - Don't tolerate change well
  - Facebook is cool

Generational Stereotypes

- Generation X
  - Independent
  - Familiar with traditional and cutting edge technology
  - Work-Life Balance

- Generation Y (Millennials)
  - Confident, ambitious
  - High expectations, sometimes unrealistic
  - Question authority
  - Concerned with diversity, societal goals, environment

Generational Stereotypes

- Generation Z
  - Financially focused
  - Competitive
  - Diversity is not even a thing, it is natural
  - Into both technology and socializing
  - Embrace change
  - This generation seems to have incorporated many of the traits of the Baby Boomers, Gen X and Millennials

The Truth about the Generations

- Every person is different, they are affected by their upbringing, natural characteristics and environment
- You can find all the traits identified in people in all generations
- Stereotyping doesn't work – there are some generational traits but they don't apply to everyone
- Listening, empathizing and accepting will work for all generations

Flaws in Research on Generational Differences

Contrary to popular narratives, the committee found, rigorous studies have shown that individuals from the same generation are just as likely to be different from one another as they are from individuals born in other generations.

The report says much of the research looking for generational differences is based on data collected at a single point in time, such as through a survey, and cannot reveal how generations' attitudes or values have changed or stayed the same. This type of research fails to distinguish between generation, period, and age effects, limiting the utility for management decisions.

Studies that have found millennials are less concerned with career advancement than with achieving work-life balance, for example, can be better explained by age differences or current work conditions. Other research has shown that the centrality of one's career declines until age 65, regardless of generation. Since the 1990s, Americans generally have tended to place greater importance on jobs that provide security, high income, and opportunities for advancement, all of which have become more difficult for workers to attain.

https://www.nationalacademies.org/news/2020/07/categorizing-workers-needs-by-generation-such-as-baby-boomers-or-millennials-is-not-supported-by-research-or-useful-for-workforce-management

Comments from Young ChE's

- You can't automatically assume that since I'm young, I'm lazy and don't want to work. This is a stereotype that drives me crazy.
- Our generation/people younger than me tend to pick jobs based on locations and opportunities for activities outside of work in addition to career fit in a role.
- In general I would rather have time off than money- a lot of younger people think more this way and would prefer to have a more flexible schedule. This doesn't mean I'm lazy, it just means I value different things more. I wish my manager/company knew this about me.

Comments from Young Engineers

- I can say that I worked with a lot of experienced engineers and all of them have been excellent mentors.
- I was also a manager of a few younger (<5 years' experience) colleagues in my previous role and this is what I find most productive and successful:
- Have older, more experienced colleagues work together with younger engineers and have experienced engineers consult on multiple projects
- Have frequent interactions between the young engineer and experienced engineer, share ideas on a whiteboard or collaboration space. Young engineers have an open mind and have lots of ideas. It's best for the experienced engineer is to have an open mind as well and work with the younger engineer to point out the risks/issues with their idea but not dismiss the idea altogether but build on it.
- The more experienced engineers should challenge the young engineer, ask them lots of questions, challenging the status quo, without directly giving the answers to make them think. Think about problem, the solution and the value it will bring compared to the next best alternative solution.

Comments from Young Engineers

- Pet peeves – when subject matter experts says it takes years of experience or it is to complex to explain/ takes years to learn. Don't withhold what you do know! I think our learning agility especially with technology, documentation etc. has greatly improved compared to the past. We don't expect to know and learn everything but there should be a framework, way of thinking, general rules of thumb etc. that can be taught or how to go and build about the knowledge. I think Albert Einstein has a quote "If you can't simplify something then you don't understand it well enough." There is something that can be taught as a start or figure out how you can get that person experience not just be a factor of time.
-
- Tips – Start putting your young talent out there to present or write publication articles etc. I feared being young and not having credibility or not knowing the answer to a question. My boss had me present and was in the audience if I needed support and also said it is okay if I don't know answers and can always say I can back to you on it. Managers and senior leaders need to find safe opportunities and be present to offer support at those opportunities to help their employees build confidence and credibility. It is important to start this early and know that the manager has to help facilitate these opportunities. AICHE is where I got my start!

Comments from Young Engineers

- . I may be an exception to the norm, since I really value what the prior generation has had to hack through in order to impart wisdom on the younger generation. So that's what I love. Prior experience to help us younger engineers avoid the pitfalls and traps of our new reality of work life after university. I have not experienced any direct ageism from older to younger in my positions, rather a great outpouring of knowledge that humbles me and encourages me for my future. I love when those who are ahead of me are excited to answer questions and provide rules of thumb and best practices.

Comments from Young Engineers

- It is often difficult for younger professionals to generate professional relationships with co-workers in an older demographic, oftentimes regardless of where people are in the "org chart." I frequently see that people tend to spend a lot more time managing relationships with people close to their age group even if those individuals are far apart in the organization because they have some common ground to discuss topics that appeal to their age demographic. While this is fine (and to some extend unavoidable and natural), people need to be aware that it can make it difficult to create broader professional relationships with people in different age groups, people who may benefit from mentoring and cultivating. Some awareness and intent in this area could go a long way.

- Constant generalizing/stereotyping of people by their generation tends to be inaccurate and unproductive, it only further divides age demographics at work. Sure, millennials or gen z people tend to be on their phones a lot or can be easily distracted, but those same traits can be found in older individuals as well. Constantly hearing about how difficult millennials are to manage gets old and simply dumping everyone in a generation into one bucket of behavior isn't appropriate.

Comments from Young Engineers

- In my opinion, the younger generation in our industry wants to be heard and wants to be taken seriously. They have worked hard to get where they are. They want to make an impact quickly and they want to do work that matters. I think they are easily frustrated by low-value work (e.g., paper pushing) and by being dismissed because they are young or because they don't have a lot of experience.

Let's All Get Along

- Diverse Workplaces are most successful and generational diversity can improve business
- See different ideas as strengths, not weaknesses
- Generations can learn from each other
- Respect, empathy, listening, teamwork
- If everyone is valued, success is likely

Conclusions

- While there may be some generational trends, generational stereotypes are not accurate
- Individual attributes are more important and all differences should be respected and seen as strengths
- Younger generations value the teachings they get from older colleagues, but they want their opinions and ideas heard as well

Contibutors

- Sadat Ulas Acikgoz
- Donna Bryant
- Meha Jha
- Tatiana Carter
- Liz Manning
- Jonathan Haling

LORI MCDOWELL

lmcdowell@mathesongas.com
281-740-2865

590958 Steps to Become a PE and What Exactly Does It Mean

Abstract:

The steps required to become a licensed engineer will be briefly reviewed. What this license means as well as what it does not will also be covered.

AIChE
The Global Home of Chemical Engineer

My Qualifying Experience/Background

- ChE with extensive experience in:
  - Environmental Engineering
  - Process Safety
  - Project, Program and Department Management
- 17 years on AIChE staff as Director of Technical Programming
- 54 years as Active AIChE Member (many jobs – BOD, Operating Councils, Division and Programming Chairs, Staff and many other assignments)
- Became a PE rather late in career
- Long-Time Member of NCEES ChE PE Exam Committee

3

AIChE
The Global Home of Chemical Engineer

Purpose of Presentation

- Introduce Session(s) - Licensing, Credentialing and Career Planning
- Explain Steps to Becoming PE
- What Does Having a PE Mean?
- What it Doesn't Mean!

4

AIChE
The Global Home of Chemical Engineer

Steps to Becoming a PE

- ✓ Education (Generally EAC/ABET Accredited)
- ✓ Experience
- ✓ Testing (FE and PE Exams)
- ✓ 55 Licensing Authorities (in US)
- ✓ NCEES - National Council of Examiners of Engineering and Surveying

Website: https:/ncees.org

5

AIChE
The Global Home of Chemical Engineer

What is a PE?

- ➢ Demonstrated Minimal Competency
- ➢ Visible Publicly (Can use PE title)
- ➢ Required Legally for Some Work
- ➢ Shows that One Subscribes to Ethical Practices
- ➢ Also Requires Maintenance of Professional Competency

6

AIChE
The Global Home of Chemical Engineer

What a PE Isn't

- Does not Demonstrate One is an Expert
- Does not Conflict with or Replace Need for Additional Credentialing

7

AIChE
The Global Home of Chemical Engineer

Remainder of Session(s)

- Will Explore Perspectives and Examples of Practicing PEs
- Will Look at Future of Credentialing
- Submit Questions in Chat Boxes and Speakers will Respond in Box

8

AIChE
The Global Home of Chemical Engineer

How to Reach Me

My e-mail is josec@aiche.org

591127 The Future of Badges, Credentials and Licenses

Abstract:

The speaker is President-Elect of AIChE, past chair of the Licensing and Professional Development Committee and is currently transitioning out of chairing the PAIC subcommittee on licensing. Ms. Grubbe will share some data around the global workforce, its mobility and how companies are now starting to fill their needs for skilled technical labor. She will also share how AIChE is responding to this trend, and will offer a view of the "career of the future" and how many of us will acquire and verify our skill sets for our future employers.

590294 Credentials, Certificates and AIChE Related Educational Programs

Abstract:

AIChE has formed both a credential for sustainability professionals and a professional certification for process safety professionals and also provides certificates for completion of AICHE Academy educational courses. All of these credential related programs have value in the workplace and career development. the differences, advantages and career utility of the programs will be reviewed in the context of professional development

590386 Control System Design Using Reinforcement Learning

Abstract:

ABSTRACT for AIChE 2020 Spring Meeting

Control System Design using Reinforcement Learning

James Cross & Emmanouil Tzorakoleftherakis
The MathWorks, Inc.

jcross@mathworks.com
617-605-5818

etzorako@mathworks.com
508-647-1759

Reinforcement learning (RL) is a technique for computing state-based decisions that lead to maximal rewards. It has much in common with optimal control theory.

As an example: *What is the schedule of control setpoints that achieves plant start-up with minimum total energy consumption?* This is the type of problem, extraordinarily complex, that RL and related techniques endeavor to solve.

In contrast to supervised machine learning methods which require training data, RL instead requires a mathematical model that describes how sequences of decisions cumulatively contribute to an overall reward. By running experiments and generating data dynamically, reward-maximizing strategies can be learned as the amount of "experience" increases. Since process simulation is a foundational aspect of plant design, such models are commonly available to the process engineer.

In this paper we focus our attention on control design. Multiple-input, multiple-output (MIMO) processes are a feature of almost all chemical plants. The design of robust control strategies is critical for maintaining consistent product quality, ensuring safe operations, minimizing downtime, and generating profit. The design process typically involves the comparative evaluation of alternative control loop configurations for interacting process units, applying domain expertise and using techniques such as the relative gain array and decouplers. *How about using RL?*

In this talk we start with a generic example to introduce the elements of RL, including agents, policies, the "environment", update algorithms, and reward functions. We then describe a MIMO process control design problem and demonstrate how RL can be used to generate a design solution. We conclude by comparing the RL results to those deriving from a traditional design approach and offering some observations about the promise of RL for broader applications.

- Abstract Text

590241 General Feature Extraction for Process Data Using Convolutional Neural Network Based Transfer Learning

Abstract:

In the past decades, data-driven methods have been widely applied in process data analytics. Particularly in process monitoring and fault detection area, many approaches were developed from simple linear models to complicated non-linear models, including PCA-based methods [1], different kernel methods [2], and various novel machine learning techniques. Nevertheless, most of these data-driven approaches are extremely specific, where the developed models can only be applied in the given process with the selected variables. In other words, a model trained on an old chemical process may not be suitable to be applied on a new process with same implementation of operation units. The lack of generalization is a major drawback of current data-driven process monitoring techniques, which potentially limits their performance on real industrial applications.

On the other hand, if we review the data in any chemical processes, a commonality can be observed that most of the measurements are pressures, temperatures, flow rates, levels and concentration of different species. Typically, their intrinsic correlations can be summarized by the first principles, such as mass balances, gas laws, energy balances and etc. Although the first principle approach requires a lot of effort to investigate a physical system, while it provides a more general solution for different kinds of systems. Additionally, commonalities also widely exist in process monitoring and fault detection models. Regardless of the number of input variables used in the model, the objective is always to correctly identify the distribution, $\sim N(\mu,\sigma^2)$ of normal operating conditions and distinguish the abnormality of test data.

With the recent development of deep learning techniques, it offers the feasibility to build models with better generalization capability. Inspired by the recent breakthroughs in computer vision field, a transfer learning approach was proposed using convolutional neural networks (CNN), in order to extract general features from process data. The proposed framework is more generalized than conventional methods, where features across different processes with different variables can be learned. The proposed CNN classifier was first trained on the famous benchmark, the Tennessee Eastman Process (TEP) [3] to learn the patterns of different kinds of faults. The features learned in the intermediate layers of the CNN classifier can be pooled out for various tasks. In this work, a fault detection method is demonstrated based on the proposed feature extraction tool, which provides a superior generalization ability that can be applied outside the TEP datasets.

References:

[1] L. H. Chiang, E. L. Russell and R. D. Braatz, Fault detection and diagnosis in industrial systems, New York City: Springer Science & Business Media, 2000.

[2] S. W. Choi and e. al., "Fault detection and identification of nonlinear processes based on kernel PCA," Chemometrics and intelligent laboratory systems, vol. 75.1, pp. 55-67, 2005.

[3] J. J. Downs and E. F. Vogel., "A plant-wide industrial process control problem," Computers & chemical engineering, vol. 17.3, pp. 245-255, 1993.

589732 Dow2Vec - How Text Embeddings Enhance NLP Models for Manufacturing

Abstract:

Over the past few years, several language embedding model architectures have been shared with the open source community which have achieved top accuracy measures across many natural language processing benchmarks. One of these techniques called Bidirectional Encoder Representations from Transformers (BERT) can provide both out-of-the-box text features for downstream NLP tasks or a fine-tuned model trained on a smaller set of labelled data. This talk will provide the audience with an overview of how to apply this model to a manufacturing-specific document classification task and share the accuracy results when compared to traditional NLP feature extraction methods.

591462 In Business No Doesn't Always Mean No, Dig Deeper

Abstract:

Have you ever wanted to ask a question, but were afraid the answer would be no? Can I have a raise, can I manage this project, Is my performance good? Or, have you ever asked a question and been told no without a reason, or when you felt the answer should have been yes? In sales, when a customer says no, do you leave and not come back? Does no mean I am on the wrong track or misinterpreted something?

If you've considered saying yes to any of those questions, then this workshop is for you. This workshop will deal with all the above and more. You will learn how to accept a no and how you might turn it to a yes. We will discuss strategies on how to make the most out of a no answer, when the answer really is no. And no can mean may things: no, forever; no, definitely not; no, not now but maybe later; no, I'm too busy to think about it; no, maybe ; or even no, but I really mean yes. We will look at different situations to determine what no really means. We will practice receiving the answer no and turning it into a positive situation.

589189 Lead It! Tools and Strategies for Effective Technical Project Leadership from a Slightly Seasoned Professional

Abstract:

Most of the young inexperienced engineers and scientist enter the workforce without grasping the reality that at some point in time in their careers they will have to take on project leadership roles. This talk will focus on technical projects and how to manage and lead them effectively. Some of the most common tools used at Dow to solve complex technically challenging process problems will be presented as well as the strategies on how to deal with people issues. Emphasis will be given to dealing with individuals stalling the projects, commonly known as NoNos. The importance of networking, recognition and project closing steps will be discussed as well.

590412 The Essential Role of Professional Chemical Engineers in Export Control

Abstract:

According to the National Society of Professional Engineers Code of Ethics, engineering has a direct and vital impact on the quality of life for all people. Indeed, the first fundamental canon is to hold paramount the safety, health, and welfare of the public. To this end, chemical engineers are essential in the U.S. Government's (USG's) chemical & biological export-control agenda, which strives to support the goals of the Geneva Protocol, Chemical Weapons Convention, and the Biological and Toxin Weapons Convention. Chemical Engineers are involved in the formulation of international export control lists with regimes such as the Australia Group, the incorporation of these lists into USG regulations, the proper adjudication of USG export-control applications, the defense of positions for this adjudication, and ultimately, the refinement and update of international norms based on past history but also the essential need to maintain pace with accelerated, emergent technology (e.g. synthetic biology). These duties require precise knowledge of the fundamentals and application of the core areas demonstrated in the attainment of a Professional Engineers' license. This presentation will demonstrate how chemical engineers are making the world a better place by using licensing measures to ensure that exports of certain chemical, biological agents, and dual-use chemical and biological manufacturing facilities and equipment, do not contribute to the spread of chemical and biological weapons.

590338 Personal Perspectives of a Practitioner on Credentialing and Licensing

Abstract:

The path we take as we develop in our careers is a unique and personal one. We begin the journey wide eyed armed with our basic education and the goal of landing our first professional job. As we move through the career our experience and skills grow, and we begin to select our path more thoughtfully. The choices to pursue certificates, credentials and/or licensure are forks in the road that help one deliberately shape a career path. This presentation will discuss differences between certificates, credentials, and licensure, and describe how each can impact a career path.

591984 Community of Mentoring

Abstract:

In multigenerational industries, institutionalizing learnings and developing the next-up technology leaders and experts are key to continued corporate stability and growth. Succession planning and mentoring are needed across all career stages for employees to reach their full value delivery capability, regardless of whether the path pursued is technical or managerial. Likewise, leveraging one's own capabilities to the fullest requires leveraging those capabilities through others, making steady development of mentoring competency a priority during all career stages. This presentation will cover Bill Docter's career track at ExxonMobil, how mentoring has been integrated throughout his career, and driving group mentoring culture as a senior technical lead in process control and optimization.

591633 A Chemical Engineer Among the Robots

Abstract:

From my first role as an Improvement Engineer to my current role leading a global robotics development program – I am often asked "How did you end up in Robotics?" I am going to share my technical career while discussing the motivations, the skills and the work that lead to the multiple jobs in my 20^{+} year career; including roles in Manufacturing, Engineering, Maintenance, Technology and now Robotics. My goal is to encourage the audience to set a vision for how they want to deliver value, build multi-disciplinary skills and create new opportunities where they least expect it.

586275 Distillation Column Explosion

Abstract:

This case study describes the investigation of an explosion that occurred during the startup of a vinyl chloride monomer (VCM) distillation column. The column had been idle for a few days while work was performed elsewhere in the process. However, the contents of the column, consisting primarily of VCM and ethylene dichloride, had not been removed prior to shutdown. A series of exothermic reactions occurred inside the column, beginning during the idle period. These reactions initially went unnoticed. However, they became more rapid and energetic as the column was started up and steam was applied. The final exotherm resulted in a failure of the column wall, releasing its contents. The contents ignited upon release, generating a large fireball and then a sustained jet fire, until the inventory was consumed. Fortunately, no one was injured as a direct result of the incident, but the column and nearby equipment were destroyed.

Following the scientific method, data from a variety of sources were reviewed and chemical and physical tests were performed to evaluate hypotheses. It was ultimately concluded that chlorine gas had leaked past a block valve intended to isolate the column from a downstream reactor vessel. Lessons learned will also be discussed.

Three Cases
Process Safety Fatalities
Better Way to Communicate Findings to Employees

Dr TJ Larkin
Larkin Communication
230 Park Avenue, 3rd Floor West
New York, New York 10169

Phone: 1-212-860-2939
Email: Larkin@Larkin.Biz
Web: www.Larkin.Biz

AIChE Spring Meeting
Session: Case Histories - GCPS Joint Session
August 19, 2020

More Communication Samples
"Lessons Learned"
PDF can be downloaded at: www.Larkin.Biz

Communication Best Practice - Lesson Learned

Nitrogen Asphyxiation During Turnaround

Employee killed. Supervisor seriously injured.

Nitrogen asphyxiation while working under a black plastic sheet.

Employee and supervisor cleaned the flange on a 48 inch pipe.

They did a black light inspection looking for any left over dirt or grease on the flange surface.

To see better in the bright sunlight, they covered the pipe opening with a black plastic sheet and climbed underneath.

The supervisor forgot he ordered a nitrogen purge the night before.

Derived from Chemical Safety Board Investigation: Report# 98-05-I-LA

Communication Best Practice - Lesson Learned

Explanation - Nitrogen Asphyxiation

During a turnaround, an oxygen feeder was removed for cleaning.

This left a 48-inch pipe opening.

employee did a black light inspection.

Black light will reveal left over dirt or grease.

In the bright sunlight, they could not see the black light results.

So, they found a black plastic sheet, covered the pipe end, and they climbed underneath.

Two crane riggers were nearby ready to lift the oxygen feeder back into service.

It was windy, so the supervisor and employee asked the riggers to hold the plastic sheet in place around the pipe opening.

After 20 minutes, the riggers became concerned because there was no talking coming from under the plastic sheet.

The riggers looked under the sheet.

The riggers found the employee:
- face down
- half inside the pipe
- his skin color was purple
- he was dead on arrival at the hospital

The riggers found the supervisor:
- unconscious
-
- facing out
- he suffered serious injuries from oxygen deprivation

Formatting Issue - Best Available Paper/Graphic

Communication Best Practice - Lesson Learned

Lessons - Nitrogen Asphyxiation

Talking Points

You Can Only Remember 4 Things Without Writing Them Down

- The supervisor forgot he ordered a nitrogen purge the night before.
- This forgetting is normal.
- Humans, on average, can only remember around 4 things in their "working memory."
- The brain can store more information in its long-term memory, but the part of the brain devoted to current "at hand" tasks (working memory) holds only about 4 items.
- As a 5th item is added, an older item is dropped out of the working memory.
- During a major turnaround, with so many demands on a supervisor, it was inevitable he would add a 5th item—the purge dropped out of his working memory.
- That's why we post signs.
- That's why we write things down.

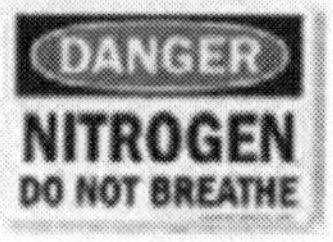

Cowan, Nelson: "The Magical Mystery Four: How is Working Memory Capacity Limited, and Why?"
https://www.ncbi.nlm.nih.gov/pmc/articles/PMC2864034/

Download more samples:

Lessons Learned Brochure

www.Larkin.Biz

This Lesson is derived from:

Chemical Safety Board Summary Report: Nitrogen Asphyxiation; Report # 98-05-I-LA.

CSB has not previewed or approved our interpretation.

Dr TJ Larkin & Sandar Larkin   Larkin Communication Consulting   www.Larkin.Biz

Communication Best Practice - Lesson Learned
Painter Dies When Scissor Truck Goes Up Unexpectedly
Derived from WorkSafeBC Incident Investigation Report 2013124710035
Dr TJ Larkin & Sandar Larkin
Larkin Communication Consulting
www.Larkin.Biz

Formatting Issue - Best Available Paper/Graphic

Communication Best Practice - Lesson Learned

What Happened - Painter Dies in Scissor Truck Accident

What they were doing:

A painter and his assistant were painting a sprinkler pipe.

They were working about 2½ feet below the ceiling.

Since the sprinkler pipe ran close to a wall, they were unable to get the truck directly under the pipe, so both workers were leaning over the top guardrail.

As they were painting the ceiling side of the sprinkler pipe, they held a mirror in one hand and paint brush in the other hand.

What happened:

The painter leaned over the top guardrail where the control panel was attached.

The painter's safety harness or perhaps overalls became entangled around the control panel joystick.

While leaning over, he unintentionally pushed the joystick forward causing the platform to go up.

The painter screamed: "What's happening?" and then, "Why is it going up?"

Since they were already working close to the ceiling, the two painters had only 5 seconds to get their bodies back inside the guardrails.

The last moments:

First, the painter attempted to untangle his clothing from the joystick.

Then, the painter tried to pull himself back inside the guardrails.

The assistant painter managed to get back inside the guardrails.

The painter didn't make it. His head became stuck between the top guardrail and the ceiling.

As his body was still pushing the joystick forward, the platform continued to rise putting ever increasing pressure on his head.

Two electricians working nearby heard the screaming and ran to the scissor truck.

They opened a compartment at the base of the truck and pushed an emergency stop button.

This caused a drop in the hydraulic pressure and slowly the platform began lowering.

The painter fell unconscious onto the platform and died a few hours later in the hospital.

The assistant painter desperately looked for the emergency stop button located on the control panel.

Finding the emergency stop button w   's body was covering the control panel.

Worse, the assistant painter was looking for a red mushroom-capped emergency stop button—but the cap was missing—only a slender pin protruded from the button casing.

The assistant painter, not trained to operate the scissor truck, did not notice the slender pin. If he had pressed the pin in, the platform would have stopped rising.

Formatting Issue - Best Available Paper/Graphic

Communication Best Practice - Lesson Learned

Primary Cause: Control Panel in Very Poor Condition

Joystick did not stick in neutral:

Normally, the joystick "sticks" in neutral. A pin on the joystick slides into a groove cut into its housing. This holds the joystick in neutral.

However, the entire control panel was covered with over-spray paint.

P ve so the joystick pin no longer slipped into the groove and held in place.

Now, the slightest pressure (around 1 kg of force) was enough to move the joystick.

Lock ring broken:

Normally, the lock ring must be raised before the joy stick is operational.

But, the lock ring was broken.

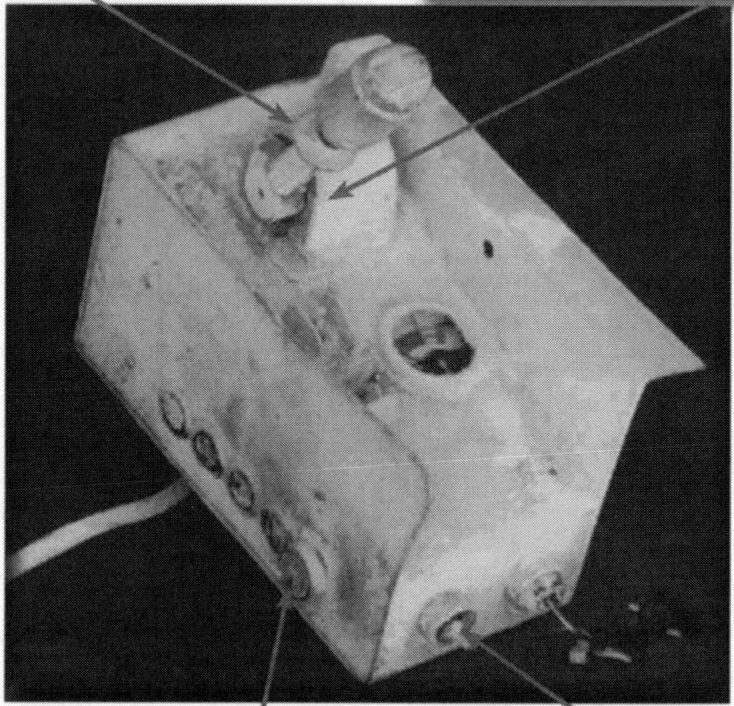

Enable switch overridden by employees:

The scissor truck design required "two-hands-on" operation.

The truck should not move, go up, or down unless an enable switch was pressed in with one hand, and the joystick moved with the aising the joystick lock ring).

Employees bypassed this feature by drilling two small holes into the enable button housing and then sliding a thin nail through the holes. This override kept the enable switch permanently pressed in.

Investigators found "two-hands-on" operation was "hated" by employees because it left them with no free hand to hold onto the guardrail when they were moving the scissor truck.

Overriding the enable switch allowed the driver to operate the joystick with one hand, and hold onto the guardrail with the other hand. Investigators found employees frequently override the enable button when using this type of scissor truck.

Emergency stop button missing its cap:

Months before this tragedy, the red mushroom-cap had fallen off the emergency stop button.

All that remained was a thin pin sticking out from the button housing.

The pin would have stopped the platform rising if it was found and pressed in.

The assistant painter did not know the cap was missing, and he was not trained to operate a scissor truck.

He frantically searched for the red mushroom-caped button and of course he never found it.

Dr TJ Larkin & Sandar Larkin   Larkin Communication Consulting   www.Larkin.Biz

Communication Best Practice - Lesson Learned

Lessons - Painter Dies in Scissor Truck Accident

Talking Points

When Defective Equipment Seems Normal

- The problems with this scissor truck's control panel were not new—they existed for months.
- Employees had grown used to these defects.
- Before beginning work, the painter completed a "Pre-Use Inspection Checklist." He did not mark anything as defective.
- These defects seemed normal to him.
- Let's talk about our equipment. What defects are we ignoring?
- What equipment defects could kill one of us today?

Overriding Safety Devices

- This scissor truck was designed for "two-hands-on" operation.
- This truck was designed to not move, go up, or down without using two hands: one hand pressing the enable button—while the other hand moves the joystick.
- Employees "hated" this two-hands-on design because it left no free hand to hold onto the guardrail, especially important when they were moving the truck to a new work location.
- So, they overrode it.
- What safety devices do we override?
- How could our overrides come back to hurt us?
- Can we get design changes made so these overrides are not necessary?

This lesson learned is derived WorkSafeBC Incident Investigation Report #2013124710035:

Scissor lift platform rises pinning worker between guardrail and ceiling

WorkSafeBC has not previewed or approved our interpretation.

Formatting Issue - Best Available Paper/Graphic

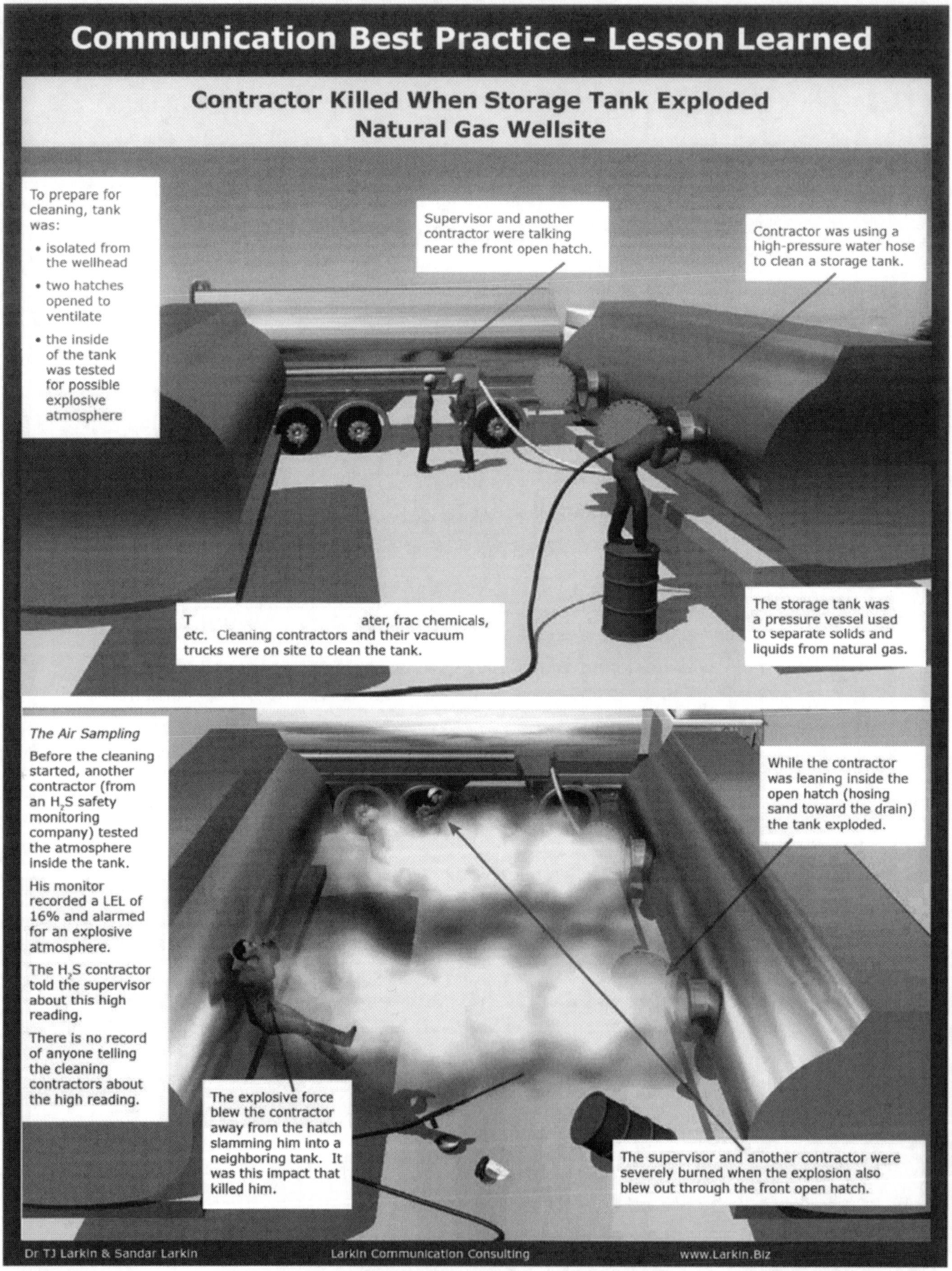

Communication Best Practice - Lesson Learned

Lessons - Contractor Killed When Storage Tank Explodes

Talking Points

When it comes to safety, are we "demanding" or "easy"?

During the incident investigation, the cleaning contractors said they followed safer procedures at other sites.

Why didn't they follow safer procedures at this site?

Because, they said: "the supervisor at this site did not demand safer procedures."

Many of us have experience at other sites.

When it comes to safety, what kind of site are we: "demanding" or "easy"?

Sampling Result Not Communicated

H_2S contractor found an explosive atmosphere inside the storage tank.

H_2S contractor told the supervisor about this dangerous result.

But, this dangerous result was not communicated to the cleaning contractors.

Worse, the dangerous result was not posted at the tank hatches.

Ignition Source Not Known

We don't know where the spark or

Two possibilities are:

1. Static electricity between the vacuum truck and tank (truck was not bonded or grounded to the tank).
2. Valv as closed, but the closed valve did not seal properly leaving a ¼ inch gap still open.

Other Serious Problems

X	no isolation	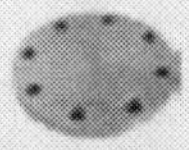	Values upstream and downstream of the tank were closed but no blinds or blanks were installed.
X	no lockout		Values were closed but no locks or tags attached.
X	no purging		Tank was not purged with nitrogen or water.
X	no grounding		The vacuum truck was not grounded or bonded by cable to the tank or the ground.

Source: This *Safety Meeting Topics* is based on an accident investigated by WorkSafeBC.

Formatting Issue - Best Available Paper/Graphic

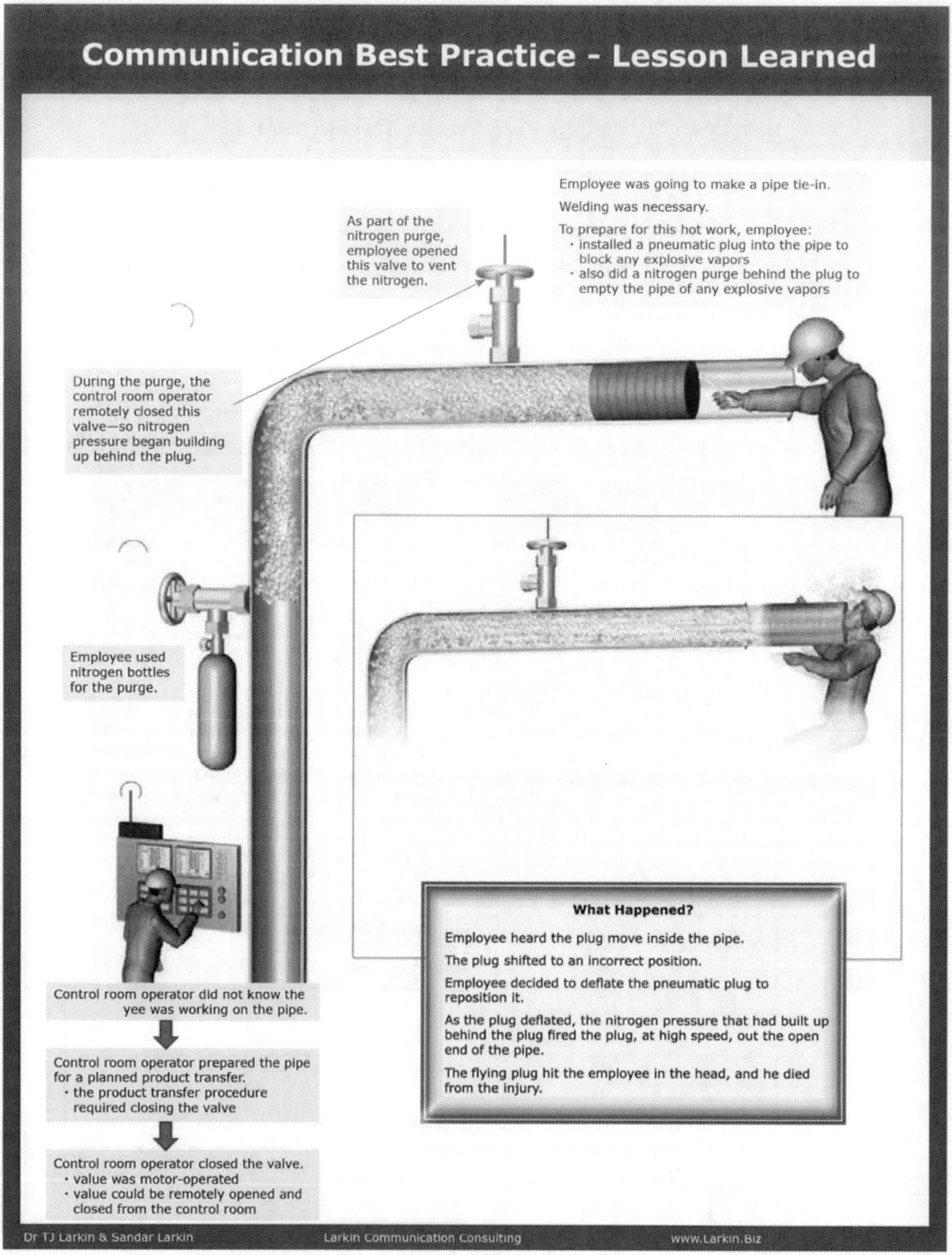

Formatting Issue - Best Available Paper/Graphic

Communication Best Practice - Lesson Learned

Lessons - Pneumatic Plug Fires from Pipe

Talking Points: Who's in the Permit Process?

Control room operators were not included in the pipe tie-in work permit.

- control room operators had no idea anyone was working on the pipe
- maintenance team did not radio the control room operators saying they were working in their area
- control room operators were preparing to move product through that pipe—that's why they closed the valve

Could this happen at our site?

Let's discuss the work we did here last week...

1. Did we communicate this work to everyone who needed to know?
2. Does our permit process include all the right people?
3. How can we improve our communication around permits?

Team Did Not Lockout the Valve

Maintenance team should have locked out the valve before beginning the purge.

If the valve was locked open, the control room operators would not have been able to remotely close the valve.

The employ
plug shifted inside the pipe.

The plug moved inside the pipe because it was under-

-
- the employ
- that's why the plug mov
  then repositioned

as designed to
withstand a backpressure up to 12 psig.

The backpressure from the nitrogen purge was estimated at only 2 to 6 psig.

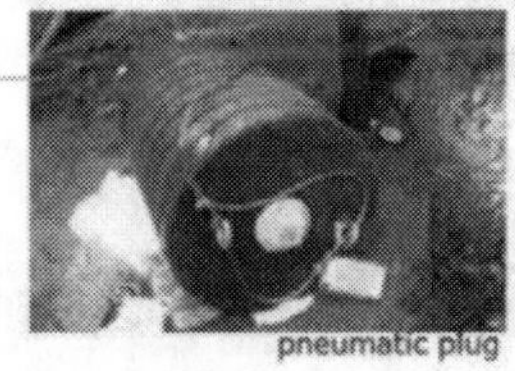

pneumatic plug

No Pressure Gauge

A pressure gauge should have been installed.

This gauge would measure any pressure buildup behind the plug.

Employees would have noticed something was wrong if they saw increasing pressure building up behind the plug.

No Barriers Blocking the Line of Fire

dangerous and should have been barricaded.

the pipe opening.

While the plug w
line-of one.

The pneumatic plug was attached to a long hose that

line-of one.

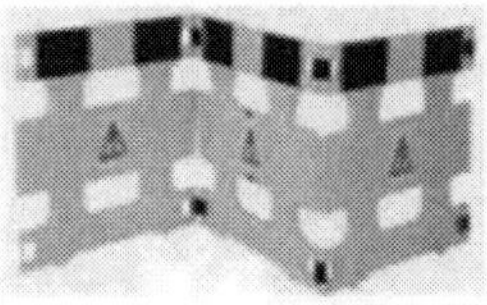

Dr TJ Larkin & Sandar Larkin Larkin Communication Consulting www.Larkin.Biz

Formatting Issue - Best Available Paper/Graphic

Larkin Communication

Email or Call Us

We would be happy to speak with you about lessons learned or any of your safety communication needs.

You may schedule a telephone call or conference call for no charge.

Phone: 1-212-860-2939; Email: Larkin@Larkin.Biz

Making a Lesson Learned

Based on your investigation, we write a *Larkin Lesson Learned* for your:

- fatality
-
- near miss

We write this lesson learned at our NYC offices; or we write this lesson learned at your location working with your team.

Working at your location, with your team, creates the most transfer of skills to your people—but where we work is your preference.

You may chose a current incident, or you may go back in time and make *Larkin Lessons Learned* for older incidents that you fear may be repeated in the future.

For fees, email our office: Larkin@Larkin.Biz

Who We Are

Dr TJ Larkin

Sandar Larkin

Since 1985, we have been helping large companies improve communication with employees.	
Book	*Communicating Change*, McGraw-Hill, New York
Most Read Paper	"Reaching and Changing Frontline Employees," *Harvard Business Review*
Newest Papers	Download our newest papers on communicating safety from our website: www.Larkin.Biz (no charge)
TJ's Background	Ph.D. in Communication (Michigan State University) M.A. in Sociology (University of Oxford)
Sandar's Background	Sandar is originally from Burma and worked with the Long Term Credit Bank of Japan before starting Larkin Communication Consulting with TJ.

Contact Information

Larkin Communication Consulting
230 Park Avenue, Suite #1000
New York, New York 10169

phone: 1-212-860-2939
email: Larkin@Larkin.Biz
web: www.Larkin.Biz

Insights to the Columbia Gas Explosions Lawrence and North Andover, MA Sept 13, 2018

Ronald J. Willey, Ph.D., PE
Dept. of Chemical Engineering, 201C CN, Northeastern University
360 Huntington Ave, Boston, MA
r.willey@northeastern.edu

Prepared for Presentation at
American Institute of Chemical Engineers
2020 Spring Meeting and 16th Global Congress on Process Safety
Houston, TX
March 30 – April 1, 2020

Insights to the Columbia Gas Explosions Lawrence and North Andover, MA Sept 13, 2018

Ronald J. Willey, PhD, PE
Dept. of Chemical Engineering, 201C CN, Northeastern University
360 Huntington Ave, Boston, MA 02115 USA
r.willey@northeastern.edu

Keywords: case history, natural gas

Abstract

On September 13, 2018 at approximately 4:00 PM EDT a series of emergency calls were received by 911 operators working the Lawrence and North Andover region of Massachusetts, USA. The calls ranged from smells of "natural gas" to actual fires to actual explosions (at least five recorded). A low pressure natural gas supply system (0.5 psig 14 inches water column) "over-pressurized" by at least one order of magnitude. Approximately 8,600 customers were affected. Of these, 131 customers had some type of "extended over pressure event" such as a release of natural gas within their basement. This is a 1.5% failure rate on systems that should have been in place when over pressure occur. Unfortunately, this 1.5% failure rate translated into a multi-million dollar loss for the gas company, and the disruption of quality of life for 8,600 residents, 100's of whom were displaced for over six months as their homes were rebuilt. One year later, full recovery is still in progress for the area. The root cause traces back to a failed "management of change" system. With proper a hazard analysis/safety review and communication, this event could have been prevented or at least mitigated.

1. Introduction

1.1 Author's note:

Much of the information in this manuscript was obtained from the NTSB final report on the incident first issued in September 2019 [1]. The author made a personal visit to the site of the incident taking photographs. Also, the author interviewed several people who have an interest in this event.

1.2 Abstract from the NTSB report:

On September 13, 2018, about 4:00 p.m. local time, a series of structure fires and explosions occurred after high-pressure natural gas was released into a low-pressure natural gas distribution system in the northeast region of the Merrimack Valley in the Commonwealth of Massachusetts. The natural gas distribution system was owned and operated by Columbia Gas of Massachusetts, a subsidiary of NiSource, Inc. Columbia Gas of Massachusetts delivers natural gas to about 325,000 customers in Massachusetts. One person was killed and 22 individuals, including three firefighters, were transported to local hospitals due to injuries; seven other firefighters incurred minor injuries. The fires and explosions damaged 131 structures, including at least 5 homes that were destroyed in the city of Lawrence and the towns of Andover and North Andover. Most of the damage occurred from fires ignited by natural gas-fueled appliances; several of the homes were destroyed by natural gas-fueled explosions. Fire departments from the three municipalities were dispatched to the fires and explosions. First responders initiated the Massachusetts fire-mobilization plan and received mutual aid from neighboring districts in Massachusetts, New Hampshire, and Maine. Emergency management officials had the electric utility shut down electrical power in the area, the state police closed local roads, and freight and passenger railroad operations in the area were suspended. Columbia Gas of Massachusetts shut down the low-pressure natural gas distribution system, affecting 10,894 customers, including some outside the area who had their service shut off as a precaution. The National Transportation Safety Board made new recommendations to the Pipeline and Hazardous Materials Safety Administration; the 31 states with an industrial exemption for natural gas infrastructure projects; the Commonwealth of Massachusetts Executive Office of Public Safety and Security; and NiSource, Inc (Ref 1. p iii)

2. Background

2.1 The Location

The location of the natural gas pipeline replacement project was in Lawrence and North Andover, Massachusetts, USA. Figure 1 A and B show the geographical location of the city. Population of Lawrence Massachusetts is approximately 76,000 and North Andover is approximately 30,500 [2] and [3]. Lawrence was one of the first industrial centers of the United States and is located along the Merrimack River.

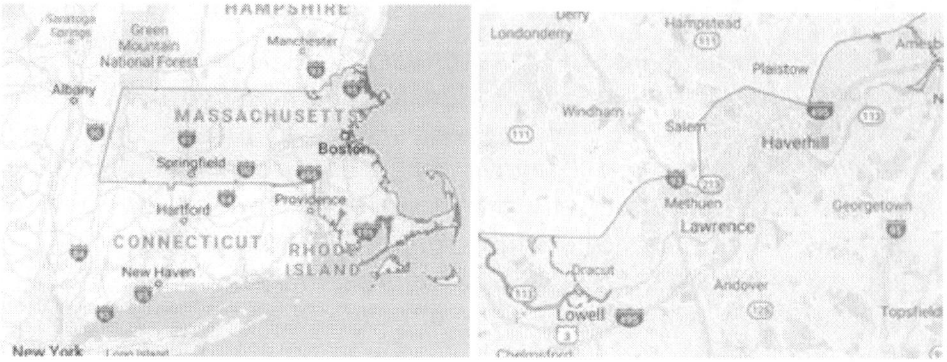

Figure 1. A and B. Geographical location of Lawrence and North Andover Massachusetts.

Shown in Figure 2 is the approximate street location where the regulator vault and sensors were located. Approximately 6 blocks of cast iron piping, dating back to the 1900's, were being replaced with new pipe polyethylene pipe.

Figure 2. This intersection is near the regulator station and sensing valves.

Figure 3. Possible location of the regulator vault, see also Figure 4.

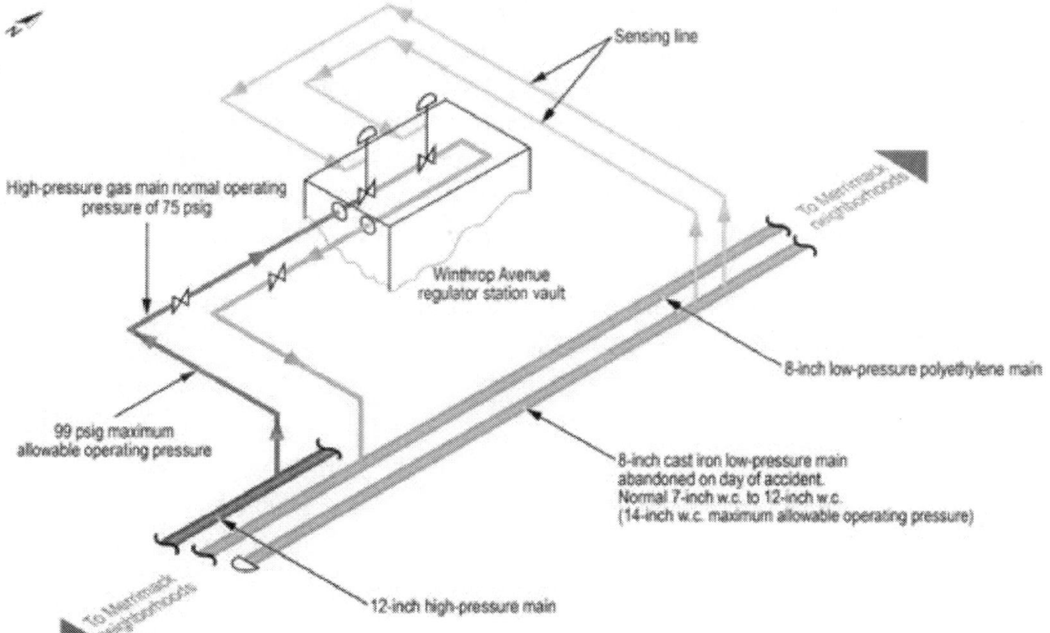

Figure 4. Schematic of the sensor, control valves, and pipe lines (Ref. 1 p 15)

2.2 The Pipeline Replacement Project

Columbia Gas of Massachusetts (CMA), a subsidy of NiSource, Merrillville, Illinois, began plans for a low pressure gas pipeline replacement project in 2016. After some delay, in 2018, Feeney Brothers Associates, a subcontractor specializing in pipe main construction, was hired to replace 7,595 feet of 90 year old of 8 inch cast-iron gas mains with 4,845 feet of polyethylene pipe. These pipelines supplied low-pressure natural gas to homes and commercial customers (ca 8,600) in the Lawrence and North Andover regions. Supply pressure ranged from 0.25 to 0.5 psi or 7 to 15 inches water column.

2.3 The Initiation

Around 4 PM, 13 September 2020, Feeney brothers closed the feed to a segment of the cast iron pipeline being replaced and brought on line new replacement pipeline. Pressure within the old fell. The sensors attached (see Figure 4) sent a signal of pressure decreasing from ca 0.5 psi to ca 0.0 psi. The regulator valves opened, and remained open, as the replacement pipe pressure continued to build beyond the 0.25 to 0.5 psi to estimates of 10 fold. The problem was that the pressure sensor was not relocated from the old pipe line. The old pipe, because it was isolated was not increasing in pressure. Yet the control system, sensing low pressure, continued to keep regulator valves open.

2.4 Fires, Explosions, and Odors

From 4 PM to 6 PM that day, due to a higher than normal pressure distribution line, localized customer gas equipment failed. Some gas appliances did not hold the increase of pressure, even at 6 psi, and the gas leaked into basements and other areas of many homes. These areas filled with natural gas. Pilots and electrical shocks provided the ignition sources. An example house destroyed in the explosion is shown in Figure 5. A chimney on the right side of house collapsed (see bricks on the right side). Leonel Rondon, 18, was killed while sitting in a car when the chimney fell upon the car. Figure 6 shows the 134 locations that has an over pressure event followed by a leak, fire, or explosion.

Figure 5. Front view of a house in which natural gas leaked to due to over pressure distribution line. (Ref 1. p 3)

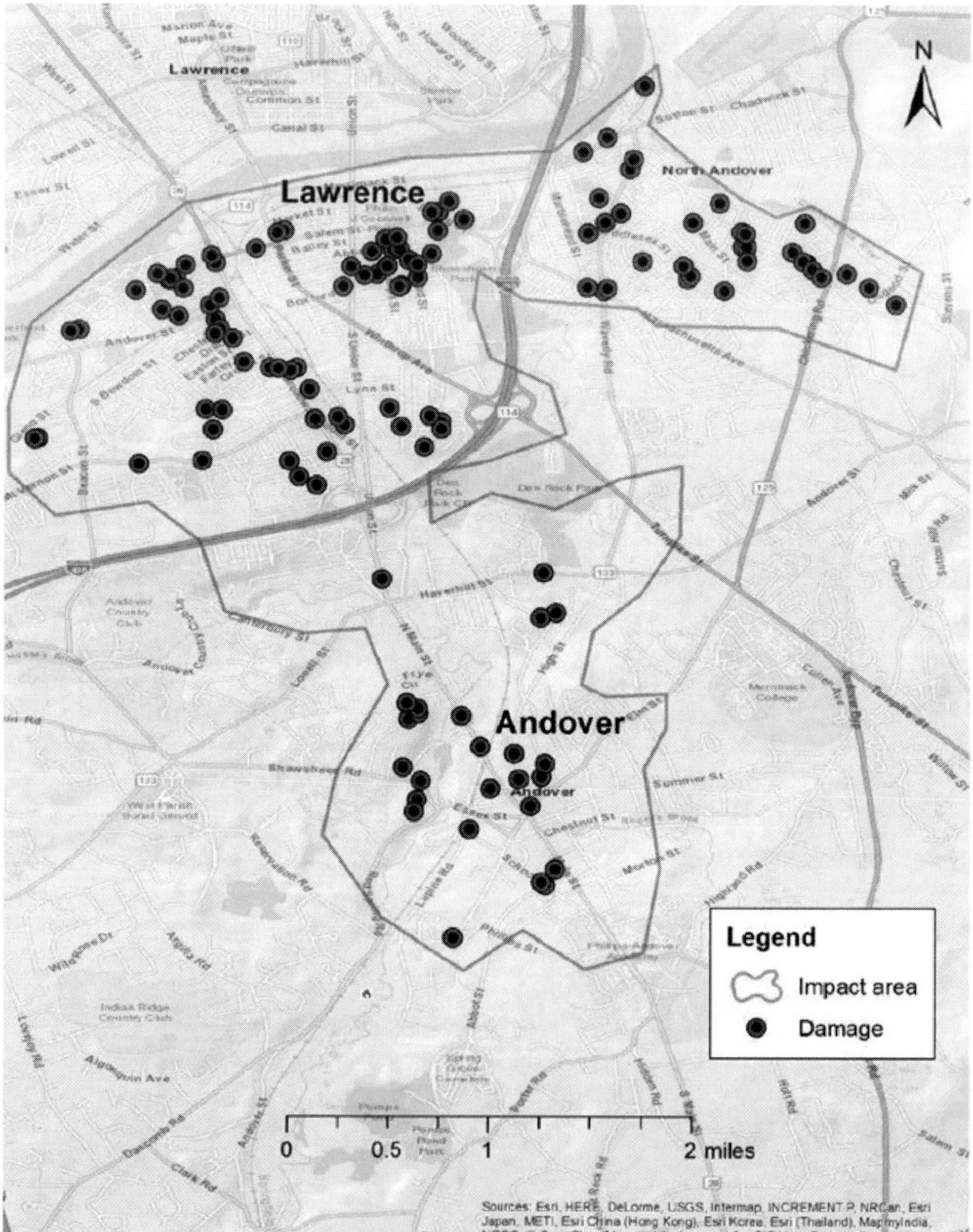

Figure 6. Map of the damaged structures in the area impacted by the overpressurization (Ref 1. p2)

2.5 Aftermath

It took nearly six months to completely re-pipe the whole area (as many as 10,000 gas customers received completely new service) and return gas service to all customers in the area. The cost is still being assessed; however, one reported settlement was at 143 million dollars, and the total losses are reported at 1 billion dollars for NiSource. [4]

3. A closer look at the layers that failed

3.1 Questionable qualifications of the Engineer-in-Charge

The engineer responsible for this project lack a Professional Engineers' License. This was possible under the industrial exemption that public utilities hold. Based on the description in the report, this person had and engineering-in-training certificate; however, the person had not completed the P. E. examination. This would indicate that the experience of this engineer is around 4 years or less.

As time moves on, the available documentation of any original project become harder to find and assemble. In this case, the 8" cast iron pipeline was installed in the early 1900s. Then, a control system consisting of a sensor and control valves were probably added around the 1970s or 80s. Then, the company changed hands with new owners.

How can we expect a young engineer to find previous information, that may have been under the previous ownership, for review in upgrade projects? Maybe, we are expecting too much? As was written in the NSTB report, (Ref 1. p 29)

> *"The documents prepared for the South Union Street project were signed by a degreed engineer who had an engineer-in-training certificate, which is held by individuals preparing to take the P.E. examination. However, he was not yet eligible to take the P.E. examination because he had not satisfied the work experience requirement."(Ref 1. P 29)*

An outcome of this oversight was this recommendation by the NSTB, nationwide. (Ref 1. p. 29)

> *On November 14, 2018, the NTSB issued Safety Recommendation P-18-5 to the Commonwealth of Massachusetts that addressed the removal of a P.E. licensure exemption for such public utility work, along with a corresponding Safety Recommendation P-18-6 issued to NiSource, the parent company of Columbia Gas of Massachusetts, recommending P.E. approval natural gas pipeline projects within NiSource (NTSB 2018). As described more fully in section 2.1, Massachusetts acted to satisfy Safety Recommendation P-18-5 less than 2 months after it was issued. Subsequent to this recommendation, the NTSB contacted two independent organizations seeking expert information on the current state of P.E. license oversight and the industrial exemptions among*

the United States and territories for major infrastructure projects. (Ref 1. p 30)

P-18-5 To the Commonwealth of Massachusetts Eliminate the professional engineer licensure ***exemption*** *for public utility work and require a professional engineer's seal on public utility engineering drawings (Ref 1. p 33)*

P-18-6 To NiSource: Revise the engineering plan and constructability review process across all your subsidiaries to ensure that all applicable departments review documents for accuracy, completeness, and correctness, and that the documents or plans be sealed by a professional engineer prior to commencing work. (Ref 1. p 34)

The Commonwealth of Massachusetts' **exemption** for the requirement of registered Professional Engineer (P.E.) to perform industrial and public utility work limits the opportunities for competently trained and experienced engineers to uncover system design and work process deficiencies. By eliminating the exemption, especially for systems involving inherently dangerous materials such as natural gas distribution systems, companies, workers, and the public are provided greater safety assurance that competent and qualified engineers, who are ethically bound to work only on projects within the scope of their expertise, will review, assess, and execute the requisite work activities according to best engineering practices and with expected safeguards. (Ref 1. p 33)

Why required a P.E.? The answer is that a qualified P.E. generally has the experience, the knowledge, and just enough hesitancy, to ask the right questions as projects go forth. They understand that protection of the public is paramount to the profession and is within a code of ethics for engineers. They also understand that old systems always have surprises. They will ask the questions such as "where is the control system and what does it consist of?" in a replacement project. When a P.E. affixes their stamp to a set of drawings, they understand that there is a code of ethics behind that stamp, a requirement that what they are stamping must withstand, to the best of their ability, safe construction and operation.

3.2 Inadequate documentation

NiSource engineering plans used during the construction work did not document the location of regulator sensing lines.

The field engineer responsible for the South Union Street Project largely relied on GIS to develop work packages. He also had access to isometric drawings containing schematics of the pipes in the regulator vaults as well as the piping and valve configurations. Sensing lines, however, were not included in the isometric drawings or GIS. Ref. 1 P 41

I note to this audience the OSHA regulations that cover our chemical industry: OSHA PSM 1910.119(d)(3)(i)(B) [5]

Piping and instrument diagrams (P&ID's); need to be up to date. However, I further advocate that the we must ask the question, and verify independently. Has someone "walked the line?" during a MOC review [6]? Have we located all of the key controllers and sensors? Even it is buried, most likely a vault will be apparent wherever sensors or regulator valves are located. Open up these vaults and inspected before the job begins. Same goes for any controllers or sensors on any lines or vessels that have to be considered in any process safety management function. We verify the information that we are working from. NiSource had over 2000 low pressure regulator runs. I wouldn't be surprised if a major oil refinery or chemical plant also have this many on any large site. This accident reminds us why we need to ask in verify important ancillary equipment that may be attached to our processes.

3.3 Management of change procedure lacking

NiSource did not perform an adequate management of change analysis for the pipeline replacement.

> *NTSB investigators found that NiSource did not use management of change (MOC) procedures for managing maintenance and construction changes to pipeline operations. The company did not conduct separate risk assessments for each construction project, critical components of a PSMS program. MOC procedures require an analysis of implications [hazard identification or process hazard analysis][1], among several other elements. Additionally, a risk identification and assessment are necessary to establish the appropriate prevention and mitigation measures to reduce the likelihood of consequences should an incident occur. CMA failed to perform such an analysis and failed to establish appropriate controls to mitigate the risks of the work that was being performed. Had NiSource adequately performed MOC, it could have immediately addressed the issue and mitigated the consequences of the event. (Ref. 1 p 35&36)*

There was a delay in the project of 1.5 years. This resulted in loss of organizational memory, specifically where the sensors were located. The safety review dated 1/16/17 was signed by engineering and construction. Measurement and Regulation (M&R) within the company did not sign off on this safety review. Sign off was not required in the original NiSource policy.

> *There was a nearly complete turnover in project personnel [between initiation and implementation]. CMA did not effectively transfer the knowledge its 2016 construction personnel had about the status of the project sensing lines to its 2018 construction personnel. (Ref. 1 p 41)*

The NTSB issued a recommendation:

[1] The bracketed statement added by Willey after review by John Murphy

Apply management of change process to all changes to adequately identify system threats that could result in a common mode failure. (P-18-8) (Ref 1. P36)

After this recommendation from the NTSB, NiSource implemented their Gas Standard 1680.010, "Tie-Ins and Tapping Pressurized Pipelines," that incorporate management of change.

3.4 Pressure regulator(s) lacking downstream pressure relief

Note the plural. That is because there are two regulators in series such that if one failed open, a second one would continue to control. Normally, these regulators must have downstream over pressure relief as specified in US Code of Federal Regulations (CFR) 49.

For example, CFR 49 section 192.195 is as follows: Protection against accidental overpressuring

. (a) General requirements. Except as provided in § 192.197, each pipeline that is connected to a gas source so that the maximum allowable operating pressure could be exceeded as the result of pressure control failure or of some other type of failure, must have pressure relieving or pressure limiting devices that meet the requirements of §§ 192.199 and 192.201. (b) Additional requirements for distribution systems [7]

In other words, at or near the regulator feeding the customer gas system exceeds a threshold supply pressure, the regulator relieves to the surroundings. An example of this detail is shown in Figure 7. Code further requires that a pressure regulator between a distribution line and a customer service must be **placed outside** with some exceptions such as a high pressure gas distribution within the building and specific gas meter rooms.

Gas meter piping supplying gas to a building at a pressure in excess of 1/2 psig (3.4 kPa gauge) shall be provided with a regulator that will reduce the pressure of the gas to 1/2 psig (3.4 kPa gauge) or less ***prior to entering*** *the gas distribution piping in the building, except where the use of higher pressure is permitted. Where gas distribution pressure in excess of 1/2 psig (3.4 kPa gauge) is permitted, it shall be regulated not to exceed the maximum pressure level as permitted by the code or the commissioner.*(NYC code [[8]

With the system 100 years old, some systems may not have had external regulators. Further, modern regulators are designed to vent should over pressure **downstream** occurs. Shown in Figure 7 is an example cut away of a regulator with overpressure ventilation capability (Itron, [9]). Three colors are shown. The red is the supply line. The blue is the downstream low pressure feed to the service. Within this specialized regulator valve is the ability of the center pressure balance to lift further should the downstream pressure exceed a limit pressure. Then, the excess gas flows through the yellow area of the valve. Generally, this is straight to the surroundings. Natural gas is lighter then air and will eventually disperse upwards from the point of release. Codes require that these regulators be placed away from any areas that may be ignition points [8].

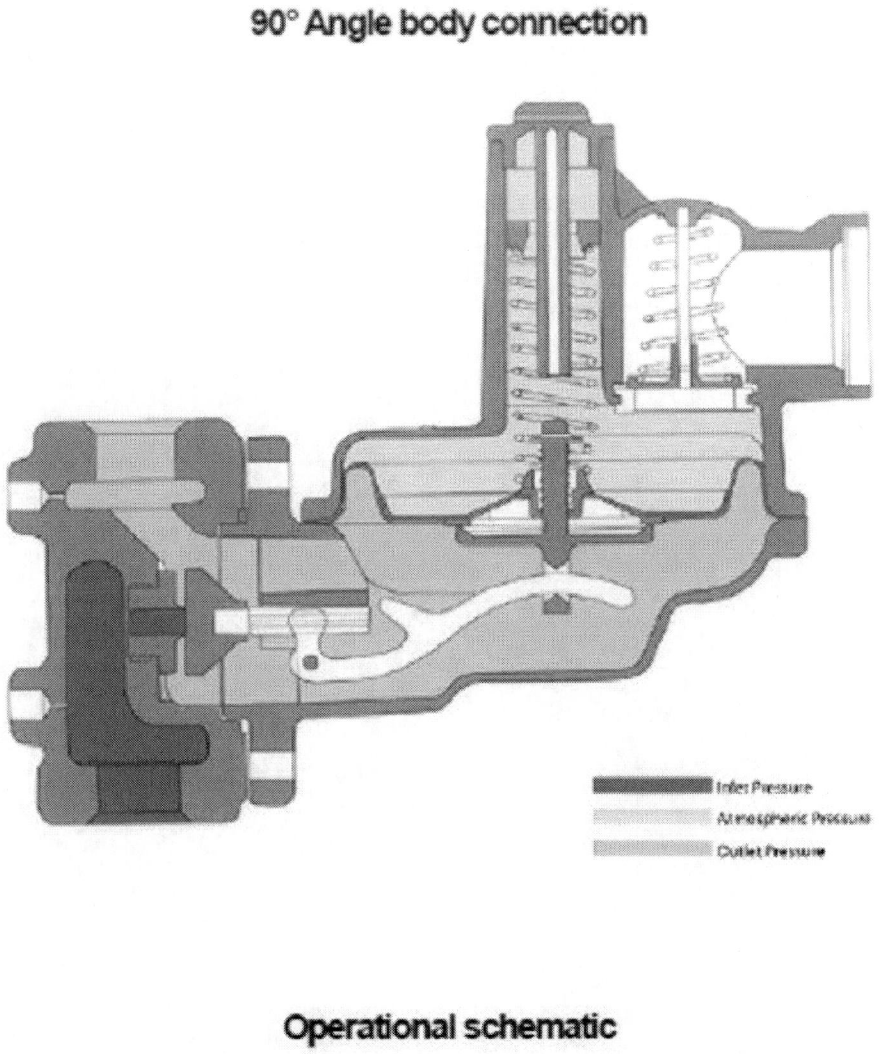

Figure 7. A typical gas regulator now required on services to prevent overpressurization. (Figure adapted from Itron, [9])

3.5 Layers of protection at the customer service point

One of the expectations of natural gas distribution systems at the customer end point is prevention of leaks within the customer's business or home should an over pressurization supply line event occurs. In this case approximately 8,450 of 8,600 customers had no leaks during the over-pressurization event. However, approximately 150 customers had an incident. Why? The following three sub section offer explanation.

3.6 Lack of, or failure, of an outside pressure relief regulator at the customer service point.

Grandfathering occurs in gas line distribution systems. Meaning that customers were connected in 1900's, others in 1910's, and others later. Early installations may not have been upgraded when codes change. This is called "grandfathering" in existing set ups. Although current codes require new connections to have over pressure protection at the customer service, older customer's may be missing this regular/relief system. For example, Figure 8 shows a customer regulator with over pressure protection. Not all services connected to the system may have had this type of regulator because they were installed before code requirement. Thus, if an over pressure event occurs, one of the layers, an outside regulator valve with over pressure relief at the customer service point, may not have existed.

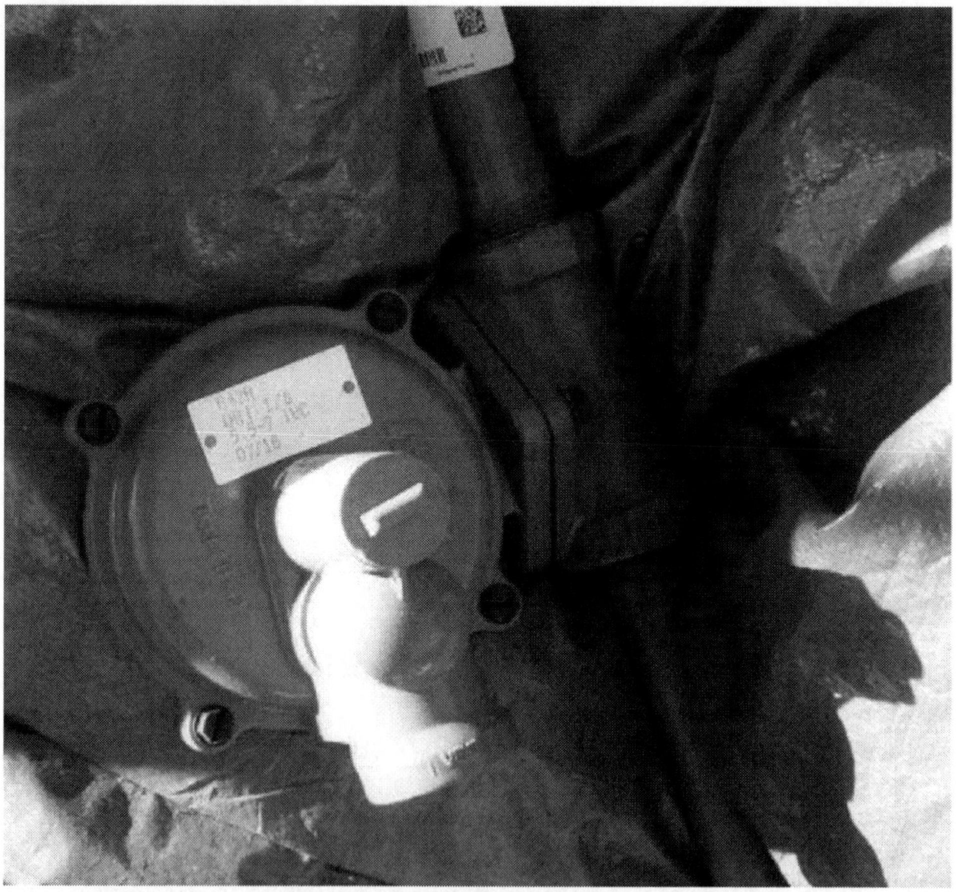

Figure 8. Example of a regulator, Itron Series B42, with excess pressure relief. Photograph by Willey taken in Lawrence shortly after the incident.

3.7 Grandfathering of internal appliances lacking over pressure protection

The national fire code, NFPA 54 Section 8.1.4 [10], has some specific requirements that new services undergo pressure tests of not less then 3 psi and be stressed by no less than 1.5 times the working pressure, holding over 30 minutes. Many appliances regulators now hold up to 25 psig or higher. Early 20th century appliances did not have this protection and may not even had a regulator between the low pressure gas supply and the appliance. These were grandfather in meaning that they were installed in the 1940's or earlier. Gas appliances with regulators came onto the market in the 60's and later. Thus, an old gas stove pilot, may be directly piped to the low pressure gas line. It can be a source of gas entering into a building. As of this writing, this detailed has not appeared in the press or any report. It is from the author's background of living in New England, and in older residences that connected to gas service many years ago.

3.8 Grandfathering of inside gas meters

Current codes and guidelines [11] require that gas meters be placed outside of a home or commercial property. Figure 8 show an example of an inside gas meter in the author's home. The gas meter was grandfathered under earlier codes and standards (ca. 1960). Presently, there is no requirement to move it outside. These meters have relief systems within to prevent total catastrophic failure. The relief, should an overpressurization event happen, goes straight into the surroundings, such as a basement.

Figure 9. An example of a gas meter inside a Northeast USA home. Newer installations places these outside. (Willey photo)

4. Conclusions and outcomes to date

4.1 Summary from the NTSB report:

In this accident, when the cast iron main with the sensing lines attached was isolated from the distribution system and abandoned in place, both regulators responded to the decreasing pressure, detected by the sensing lines, by fully opening. Both regulators were disabled simultaneously by the single event of isolating the cast iron main, which eliminated the redundancy of using dual regulators. In this accident and the earlier accidents discussed above, the overpressure occurred as the result of a single failure. In engineering analyses, such a situation is referred to as a common mode failure. Therefore, the NTSB concludes that the multiple overpressurization accidents investigated by the NTSB over the past 50 years demonstrate that low-pressure natural gas distribution systems that use only sensing lines and regulators as the means to detect and prevent overpressurization are not optimal to prevent overpressurization accidents. Thus, the NTSB recommends that PHMSA revise 49 CFR Part 192 to require overpressure protection for low-pressure natural gas distribution systems that cannot be defeated by a single operator error or equipment failure.

Less than 2 months after the safety recommendation was issued, on December 28, 2018, Bill H.5005, requiring that licensed P.E.s review and approve engineering plans developed by or on behalf of natural gas companies, to ensure the safe construction, operation, and maintenance of natural gas infrastructure, was passed by the Massachusetts House of Representatives. The act applies to engineering work or services on natural gas distribution systems that could pose a material risk to public safety, as determined by the DPU, performed by or on behalf of a natural gas company. Moreover, the act requires any engineering plans or specifications for engineering work or services that could pose a material risk to public safety, developed by or on behalf of a natural gas company, to bear the stamp of approval of a licensed P.E.39 After the Massachusetts Senate passed the act, it was signed by the governor on December 31, 2018, as Chapter 339 of the Acts of 2018. (Ref. 1 p 33).

4.2 The value of process safety as a part of the undergradate chemical engineering curriculum.

The author is reminded about recommendations from the US Chemical Safety Board from the T-2 Accident:[12]

American Institute of Chemical Engineers
2008-03-I-FL-R1

Work with the Accreditation Board for Engineering and Technology, Inc. to add reactive hazard awareness to baccalaureate chemical engineering curricula requirements.

2008-03-I-FL-R2

Inform all student members about the Process Safety Certificate Program and encourage program participation.

Accreditation Board for Engineering and Technology, Inc.

2008-03-I-FL-R3

Work with the American Institute of Chemical Engineers to add reactive hazard awareness to baccalaureate chemical engineering curricula requirements

This recommendation should be extended to mechanical engineers who train for pipe design and process design type positions.

For the AICHE audience keep in mind that Process Safety certification CCPSC is available from the CCPS. Another resource for process safety engineering to consult is the CCPS Risk Based Management Book (the "Red Book) [13].Another suggestion/recommendation is to extend OSHA PSM to cover utilities handling flammable gases and liquids. Lastly, if buildings are utilizing natural gas, should flammable gas detectors be installed?

5. Acknowledgements

The author acknowledges partial financial support for this his investigation of the accident from Practical Applications, Waltham, Massachusetts. Mr. John Murphy is also acknowledged in reading the draft manuscript and provide further suggestions for improvements. Helpful discussions with Dr. Laurence Britton is also much appreciated.

6. References:

[1] National Transportation Safety Board, Pipeline Accident Report Overpressurization of Natural Gas Distribution System, Explosions, and Fires in Merrimack Valley, Massachusetts September 13, 2018, (2019), 73, https://www.ntsb.gov/investigations/AccidentReports/Pages/PAR1902.aspx.

[2] Population of Lawrence Massachusetts 2020, https://suburbanstats.org/population/massachusetts/how-many-people-live-in-lawrence, Accessed, 1/21/2020

[3] North Andover, Massachusetts Demographics Data, https://www.towncharts.com/Massachusetts/Demographics/North-Andover-town-MA-Demographics-data.html, Accessed, 4 Feb 2020

[4] Columbia Gas reaches $143 million settlement for Merrimack Valley explosions, https://www.boston.com/news/local-news/2019/07/29/columbia-gas-settlement-merrimack-valley-explosions, Accessed, 4 Feb 2020

[5] OSHA, 1910.119 - Process safety management of highly hazardous chemicals., https://www.osha.gov/laws-regs/regulations/standardnumber/1910/1910.119, Accessed, 13 Feb 2020

[6] J. J. Forest, Walk the Line, *Process Safety Progress,* 34 (2015), 126-129, https://aiche.onlinelibrary.wiley.com/doi/abs/10.1002/prs.11724.

[7] United States Government Printing Office, Part 192 Transportation of natural and other gas by pipeline: Minimum Federal safety standards,

[8] NYC Fuel Gas Code 2014 > E Meters and Gas Service Piping > E.2 Gas Regulator and Gas Regulator Vent Outlets https://up.codes/s/gas-regulator-and-gas-regulator-vent-outlets, Accessed, 7 Feb 2020

[9] Itron, B42 Series Residential Regulator, http://deaconind.com/Customer/DEINSU/specpages/B42.pdf, Accessed, 7 Feb 2020

[10] National Fuel Gas Code, National Fire Protection Association and the American Gas Association, NFPA 54-2018, ANSI Z223.1-2018, (2018)

[11] InspectAPedia: Gas Meter & Piping Clearance Distance Codes & Specifications, https://inspectapedia.com/plumbing/Gas_Meter_Clearances.php#Codes, Accessed, 7 Feb 2020

[12] T2 Laboratories Inc. Reactive Chemical Explosion https://www.csb.gov/t2-laboratories-inc-reactive-chemical-explosion/, Accessed, 13 Feb 2020

[13] Center for Chemical Process Safety, Guidelines for Risk Based Process Safety, John Wiley & Sons, (2007)

583228 How Stressed Is Your Facility? Past Events Resulting from Structural Stress

Abstract:

When we think about process safety, we tend to focus on the process related elements, such as flow, temperature, pressure etc. However, a key element that is very important is the physical structures that thank form part of or support the process. How stress is applied to structures or equipment can have a significant effect on the process safety outcomes. A well know process safety incident that involved a stress related failure was the Flixborough explosion, where the stresses in the temporary pipe were not considered. Another is the structural failure in the Formosa Point Comfort fire, where structural steel lacked passive fire protection, and was not able to withstand its load in a fire scenario. Failures due to stress loading are not limited to the processing industries, for example in 1970 the Westgate Bridge in Victoria Australia collapsed during construction due to stress loads in in the box girder. Similarly, in 1969 and 1970 two other box girder bridges collapsed during construction. Stress related failures can also be seen in liquid tankers that 'break their back' during loading or unloading, atmospheric storage tanks that implode on discharge due to the vacuum pressure or the in-flight break-up incidents that occurred with the de Havilland Comet. All these types of failures are related to the structural stresses, though the reason for the stress may be different. This paper will explore some case studies of stress related failures highlighting possible learnings for the processing sector in identifying and managing structural stress.

584881 The Phillips Petroleum Explosion Thirty Years Later: Remembering the Lessons from This Incident Today

Abstract:

October 23, 2019 marked the thirty-year anniversary of a devastating fire and explosion that claimed the lives of 23 workers at a major chemical plant in Pasadena, Texas. That explosion, which occurred at the Phillips 66 Houston Chemical Complex (HCC) on the Houston Ship Channel, helped modernize process safety and risk management programs in the United States. However, industry's pursuit of safe process design and operation started long before the explosion in 1989. In addition, process safety and risk management thirty years ago was much different from what it is today. For example, in 1989 there were no process safety engineer roles in industrial divisions. Neither were there any process safety coordinators directing site compliance activities for process safety principles or regulations. In fact, the United States regulation for Process Safety Management (PSM) was under development in 2989 and was not issued until 1992.

Unfortunately, significant incidents resulting in fatalities, injuries, environmental damage, and property loss have continued over the last thirty years. The refinery fire and explosion in Texas City, Texas (2005), the offshore rig fire, explosion, and oil spill in the Gulf of Mexico (2010), and the toxic chemical release in LaPorte, Texas (2014) are grim reminders that global process safety improvement efforts have yet to achieve the goal of preventing or minimizing the consequences of catastrophic releases of toxic, reactive, flammable or explosive chemicals.

Many employees entering the industrial workforce today are unaware of how behaviors cultivated in a production environment can inadvertently undermine the performance of a process safety program—as they did in 1989 and continue to do so today. Recognizing and changing such behaviors requires understanding not only what happened, but more importantly how and why incidents like this explosion occurred. This level of understanding makes it possible to apply experience from the past to the systems we control today.

This paper describes the operating process, its complexities, and the maintenance activity which resulted in the explosion. Since the incident findings from the 1989 explosion apply today, this paper explores behaviors in today's plants which have and could ultimately lead to more incidents if not recognized and addressed proactively.

A common issue between 1989 and today explored in this paper is the administrative control for managing maintenance-related workaround procedures on existing engineering designs. Lessons learned then that apply now include the following: adhering to safe work practices; complying with standards, policies, and regulations; managing the removal of or disabling of safeguards; and strong operational discipline across the organization (in this case, by personnel operating and maintaining the equipment).

If we are to reduce injuries, fatalities, environmental harm, and property damage, we must continue to remind ourselves why we must keep constant vigilance on our process safety risk reduction efforts each and every day.

587051 Have We Learned Our Lessons: Revisiting the 2004 Skikda Explosion 16 Years Later

Abstract:

On January 19, 2004, at approximately 6:40 PM, a massive explosion and ensuing occurred at a liquefied natural gas (LNG) train in Skikda, Algeria. The massive blast killed 27 people and caused significant damage to the facility More specifically, the explosion disabled three LNG lines, which impacted approximately 2% of the world's liquefaction capacity. The explosion was the result of a large hydrocarbon leak that was ingested into the facility's main boiler firebox, where it ignited and breached the boiler. The boiler explosion then ignited the extensive vapor cloud that formed around the train and led to the significantly larger explosion.

This paper will first present the cascading effects which led to the cause and origin of the catastrophic explosion event, which included: (1) large hydrocarbon leak; (2) vapor cloud formation and ingestion into the boiler firebox via the air inlet; (3) the initial boiler explosion that subsequently ignited the extensive vapor cloud, resulting in very high explosion pressures. Next, this paper will discuss key lessons learned on how such an event could have been avoided by the use of necessary mitigation measures that could have been applied.

AUTHOR INDEX

AIChE
120 Wall Street, FL 23
New York, NY 10005-4020

CURRAN ASSOCIATES INC.
proceedings
.com

ISBN 978-1-7138-2347-6
9 781713 823476